PROBLEMS
IN
OPTICS AND SOUND

PROBLEMS IN OPTICS AND SOUND

M. G. RAJ

ANMOL PUBLICATIONS PVT. LTD.
NEW DELHI-110 002 (INDIA)

ANMOL PUBLICATIONS PVT. LTD.
4374/4B, Ansari Road, Daryaganj
New Delhi-110 002

Problems in Optics and Sound

First Edition 1998
ISBN 81-261-0052-4

PRINTED IN INDIA

Published by J.L. Kumar for Anmol Publications Pvt. Ltd., New Delhi-110 002 and Printed at Mehra Offset Press, Delhi.

Contents

Preface

'Optics' and 'sound' are important branches of Physics. In the present book basic theories of optics and sound are explained with the help of experimental numerical problems.

Students, researchers and teachers will find this book most useful and informative.

I am grateful to my learned friends, who has helped me in one way or another in preparation of this book. I am also grateful to Shri J. L. Kumar for undertaking publication of this book.

—M.G. Raj

Preface

Optics and sound are important branches of Physics. In the present book basic theories of optics and sound are explained with the help of experimental/numerical problems.

Students, researchers and teachers will find this book very useful and informative.

I am grateful to my learned friends who has helped me in one way or another in preparation of this book. I am also grateful to Shri J. L. Kumar for undertaking publication of this book.

—M.C. [illegible]

1

Harmonic Motion

Problem 1. An automobile moves with a constant speed of 50 mi/hr around a track of 1 mi diameter. What is the angular velocity and the period of the motion?

Fig.

Solution: For circular motion, the angular velocity ω, the radius r, and the linear velocity v obey the relation $\vec{v} = \vec{w} \times \vec{r}$ as shown in the figure. Since ω and r are perpendicular to each other, this reduces to $v = \omega r$.-

$$\omega = \frac{v}{r} = \frac{50\, mi/hr}{0.5 mi} = 100 \text{ rad/hr}$$

The period τ is the time duration of one complete cycle of motion around the circular path. In linear motion, $x = vt$. This equation can be applied to circular motion with linear velocity v replaced by ω and linear distance x replaced by θ expressed in radians. In one cycle of motion, the automobile travels 2π radians. Therefore $\omega t = \theta \quad t = \theta/\omega \quad \tau = 2\pi/\omega$

$$\tau = \frac{2\pi}{\omega} = \frac{2\pi}{100\, rad/hr} = 0.063 \text{ hr} = 3.8 \text{ min.}$$

Problem 2. A uniform circular disk of radius 25 cm is pivoted at a point 20 cm from its centre and allowed to oscillate in a vertical plane under the action of gravity. What is the period of its small oscillations? Take g as π^2 m . s^{-2}.

Solution: The moment of inertia of a uniform circular disk of radius R and mass M about an axis through its centre perpendicular to its plane is ½ MR^2. By the parallel-axes theorem, the moment of inertia about a parallel axis a distance h from the first is $I = \frac{1}{2}MR^2 + Mh^2$.

The disk is acting as a physical pendulum; and hence its period for small oscillations is given by

$$T = 2\pi\sqrt{\frac{I}{Mgh}}$$

where I is the moment of inertia of the physical pendulum about its axis of suspension. Hence,

$$T = 2\pi\sqrt{\frac{\frac{1}{2}MR^2 + Mh^2}{Mgh}}$$

$$= 2\pi\sqrt{\frac{\frac{1}{2}\times 0.25^2\,m^2 + 0.2^2\,m^2}{\pi^2\,m\,.\,s^{-2}\times 0.2m}}$$

$$= 2\sqrt{0.35625\,s} = 1.193\text{ s.}$$

Problem 3. Suppose that a mass of 8 grams is attached to a spring that requires a force of 1000 dynes to extend it to a length 5 cm greater than its natural length. What is the period of the simple harmonic motion of such a system?

Solution: An interesting property of springs is that the length that they stretch is directly proportional to the applied force. The magnitude of this force is:

$$F = kx$$

where k is the force constant. The force constant is

$$k = \frac{F}{x} = \frac{1000\,dynes}{5cm}$$

$$= 200\text{ dynes/cm}$$

Therefore, the period is by definition:

$$\tau = 2\pi \sqrt{\frac{m}{k}}$$

$$= 2\pi \sqrt{\frac{8g}{200\ dynes/cm}}$$

$$= 2\pi \sqrt{\frac{8g}{200\ g\text{–}cm/cm\text{–}sec^2}}$$

$$= 2\pi \times \sqrt{\frac{4}{100}\ sec^2}$$

$$= 2\pi \times 0.2 \text{ sec}$$

$$= 1.26 \text{ sec}$$

and the frequency is by definition:

$$\nu = \frac{1}{\tau} = \frac{1}{1.26\ sec}$$

$$= 0.8 \text{ Hz.}$$

Problem 4. A 5.0–1b ball is fastened to the end of a flat spring. A force of 2.0 lb is sufficient to pull the ball 6.0 in. to one side. Find the force constant and the period of vibration.

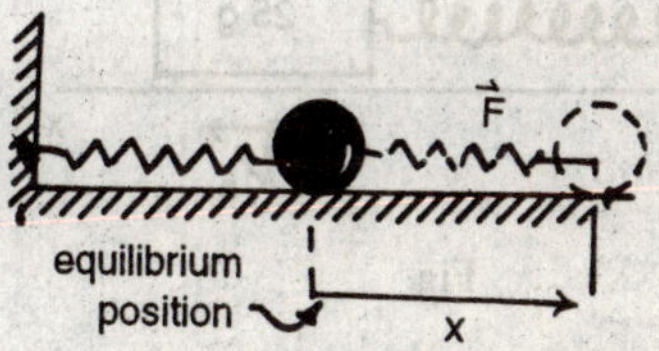

Fig.

Solution: The restoring force on the ball as it is displaced from the equilibrium point 0 (see the figure) is, by Hooke's law,

$$\vec{F} = -k\vec{x}$$

where x is the displacement and k is the spring constant. Hence,

$$-\kappa = \frac{|\vec{F}|}{|\vec{x}|} = \frac{2.0\ 1b}{0.50 ft} = 4.0\ \text{lb/ft}$$

The mass of the ball is

$$m = \frac{w}{g} \text{ where } w \text{ is the ball's weight. Hence,}$$

$$m = \frac{5.0\ lb}{32\ ft/sec^2} = 0.16\ \text{slug}$$

The period of oscillation is given by

$$T = 2\pi\sqrt{\frac{m}{k}} = 2\pi\sqrt{\frac{0.16\ slug}{4.0\ lb/ft}} = 1.2\ \text{sec.}$$

Problem 5. Let the mass of the body in the diagram be 25 gm, the force constant k be 400 dynes/cm, and let the motion be started at $t = 0$ by displacing the body 10 cm to the right of its equilibrium position and imparting to it a velocity toward the right of 40 cm/sec. Compute (a) the period T, (b) the frequency f, (c) the angular frequency w, (d) the total energy E, (e) the amplitude A, (f) the phase angle θ_0, (g) the maximum velocity v_{max}, (h) the maximum acceleration a_{max}, (j) the coordinate, velocity, and acceleration at a time $\pi/8$ sec after the start of the motion.

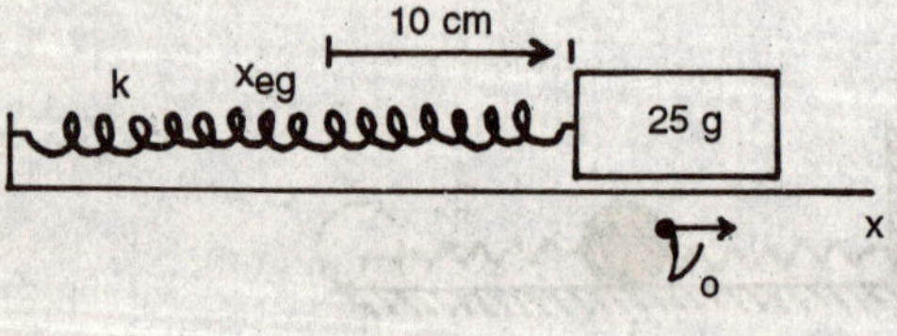

Fig.

Solution: Note that the period is independent of amplitude or by implication, the initial velocity.

$$\text{(a) } T = 2\pi\sqrt{\frac{m}{k}} = 2\pi\sqrt{\frac{25gm}{400\ dynes/cm}}$$

$$= \frac{\pi}{2}\ \text{sec} = 1.57\ \text{sec.}$$

(b) $f = \frac{1}{t} = \frac{2}{\pi}\ \frac{vib}{sec} = 0.638\ \frac{vib}{sec}$

(c) $\omega = 2\pi f = 4$ rad/sec

(d) The total energy, (that is, the sum of potential and kinetic energies), stays constant as a spring oscillates. To find the spring's energy, we will compute its energy at time $t = 0$. The object has both kinetic and potential energy. Its kinetic energy is $\frac{1}{2} mv_0^2$, with m, the mass of the body, and v_0, its initial velocity. Its potential energy is $\frac{1}{2} kx_0^2$, where x_0 is the object's distance from equilibrium, and k is the spring constant. Then

$$E_{total} = \tfrac{1}{2} mv_0^2 + \tfrac{1}{2}\ kx_0^2$$

$$= \tfrac{1}{2}\ (25\ \text{gm})\ (40\ \text{cm/sec})^2 + \tfrac{1}{2}\ (400\ \text{dynes/cm})\ (10\ \text{cm})^2$$

$$= 20{,}000\ \text{ergs} + 20{,}000\ \text{ergs} = 40{,}000\ \text{ergs}$$

(e) The amplitude of oscillation is the distance between the equilibrium position x_{eq} and the point of maximum extension. To find the maximum extension we find the position at which the system's mechanical energy of 40,000 ergs if completely potential. This occurs when the kinetic energy is zero. Then

$$\tfrac{1}{2} kA^2 = 40{,}000\ \text{ergs}$$

$$\tfrac{1}{2}\ (400\ \text{dynes/cm})\ A^2 = 40{,}000\ \text{ergs}$$

$$A^2 = 200\ \text{cm}^2; \qquad A = 10\sqrt{2}\ \text{cm}$$

(f) If x_0 is the displacement and A the amplitude, the initial angular displacement or phase angle θ_0 is defined by:

$$\sin\theta_0 = x_0/A = 1/\sqrt{2}, \quad \theta_0 = \pi/4\ \text{rad.}$$

(g) To obtain v_{max} we find the point at which the energy is all kinetic. This occurs at the equilibrium point, $x = 0$. Then

$$E = KE + PE = KE + 0 = \tfrac{1}{2}\ mv^2_{max}.$$

$$\tfrac{1}{2}\ mv^2_{max} = 40{,}000\ \text{ergs}$$

$$\tfrac{1}{2}\,(25\text{ gm})\,(v^2_{max}) = 40{,}000\text{ ergs}$$

$$v^2_{max} = 3{,}200\text{ cm}^2/\text{sec}^2$$

$$v_{max} = 40\sqrt{2}\text{ cm/sec}$$

(h) The maximum acceleration occurs at the ends of the path where the force is a maximum. This force is given by Hooke's Law, $F = -kx$, where k is the spring constant and x is the displacement of the spring from equilibrium. At maximum extension, the displacement from equilibrium is just the amplitude. Hence,

$$F_{max} = -kx_{max} = -kA = -(400\text{ dynes/cm})\,(10\sqrt{2}\text{ cm})$$

$$= -4000\sqrt{2}\text{ dynes}$$

Forces produce accelerations according to the law $F = ma$. Therefore

$$F_{max} = -4000\sqrt{2}\text{ dynes} = ma_{max} = (25\text{ gm})\,(a)$$

$$a_{max} = -160\sqrt{2}\text{ cm/sec}^2$$

At maximum extension, the acceleration is greatest and in the direction of the equilibrium point of the spring and hence in a direction which we define as negative. The motion of an oscillating spring is highly symmetric and at the point of maximum compression, the acceleration will again reach this maximum but will this time be in the positive direction.

(j) The equation for the object's position is

$$x = A\sin(\omega t + \theta_0)$$

where A is the amplitude, ω is the angular velocity, t is the time variable, and θ_0 is the initial angular displacement. The velocity, v, and acceleration, a, are found from

$$v = \frac{dx}{dt}; \qquad a = \frac{dv}{dt} = \frac{d^2x}{dt^2}$$

We then have

$$x = 10\sqrt{2}\sin\left(4t + \frac{\pi}{4}\right),$$

$$v = 40\sqrt{2}\cos\left(4t + \frac{\pi}{4}\right)$$

$$a = -160\sqrt{2}\sin\left(4t + \frac{\pi}{4}\right)$$

When $t = (\pi/8)$ sec, the phase angle is

$$\left(4t + \frac{\pi}{4}\right) = \frac{3\pi}{4} \text{ rad,}$$

$$x = 10\sqrt{2}\sin(3\pi/4) = 10 \text{ cm,}$$

$$v = 40\sqrt{2}\cos(3\pi/4) = -40 \text{ cm/sec,}$$

$$a = -160\sqrt{2}\sin(3\pi/4) = -160 \text{ cm/sec}^2.$$

Problem 6. A simple pendulum consists of a mass *m* hung on the end of a string of length L. Find the natural frequency for small oscillations.

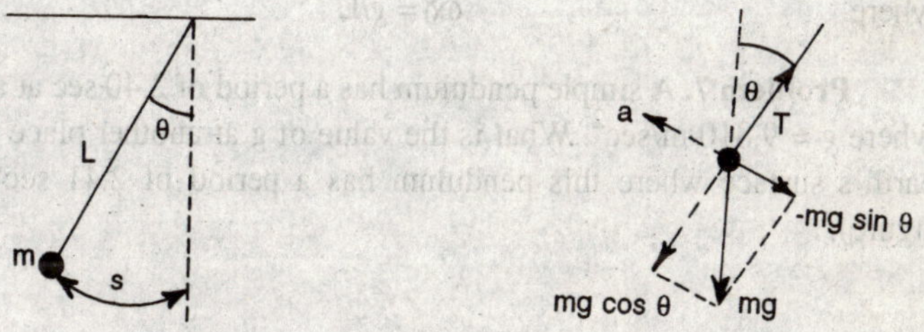

Fig.

Solution: We start by drawing a diagram of the forces acting on the mass *m*. The restoring force in the direction of motion is $-mg \sin\theta$. Thus the equation of motion is

$$ma = -mg\sin\theta$$

Now we can suppose that θ is small so that we can make the approximation $\sin\theta = \theta$. This is accurate to better than 1 per cent for $\theta = 15°$ and is better than 5 per cent for $\theta = 30°$.

The displacement of the mass is given by the arc *s*.

Thus $$s = L\theta$$

$$v = \Delta s/\Delta t = L\Delta q/\Delta t = L\omega$$

where $\omega = \dfrac{\Delta\theta}{\Delta t}$ = angular velocity. And the acceleration a is given by

$$a = \Delta v/\Delta t = L\Delta w/\Delta t = L\alpha$$

where $\alpha = \dfrac{\Delta\omega}{\Delta t}$ = angular acceleration. Then finally the equation of motion reduced to

or $$mL\alpha = -mg\theta$$

$$\alpha + (g/L)\,\theta = 0$$

The angular acceleration α is proportional to the negative of the angular displacement θ, so that the motion is simple harmonic with natural angular frequency ω_0 given by

$$\alpha + \omega_0^2\,\theta = 0$$

where $$\omega_0^2 = g/L$$

Problem 7. A simple pendulum has a period of 2.40 sec at a place where $g = 9.810$ m/sec^2. What is the value of g at another place on the earth's surface where this pendulum has a period of 2.41 sec? (See figure).

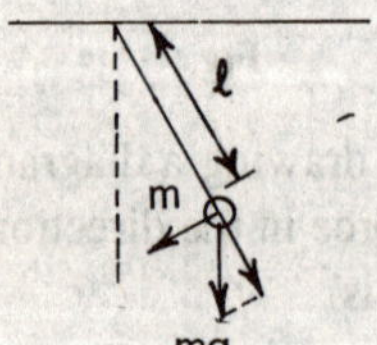

Fig.

Solution: The period of oscillation is given by

$$T = 2\pi\sqrt{\frac{l}{g}}\ .$$

Therefore, two different periods, T_1 and T_2, will correspond to two different gravitational constants, g_1 and g_2, as follows:

$$\frac{T_1}{T_2} = \frac{2\pi\sqrt{\frac{l}{g_1}}}{2\pi\sqrt{\frac{l}{g_2}}} = \sqrt{\frac{g_2}{g_1}}$$

or

$$\frac{2.40}{2.41} = \sqrt{\frac{g_2}{9.810 m/sec^2}}$$

$$g^2 = 9.739 \text{ m/sec}^2$$

Problem 8. Calculate the frequency of oscillation of the configuration shown in the figure. All surfaces are frictionless.

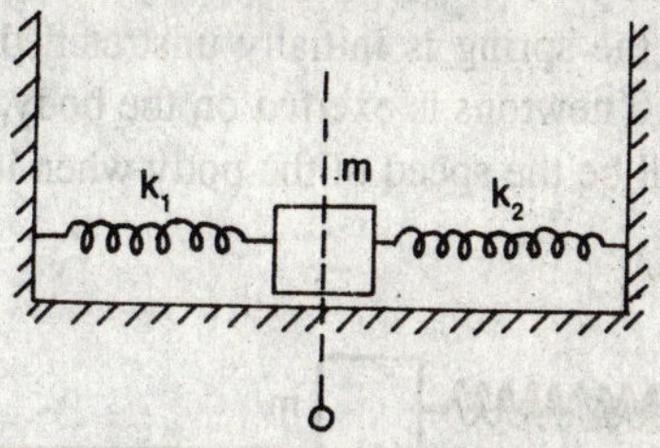

Fig.

Solution: When the mass *m* moves, one spring is always strectched, and the other is always compressed by the same length. Thus:

$$\Delta x_1 = -\Delta x_2$$

where Δx_1 is the distance that the spring with force constant k_1 stretches, and Δx_2 that of the other spring. Note that a negative stretching distance representsa distance compressed. We denote the distance of the mass to the right of the origin 0 by Δx. Thus:

$$\Delta x_1 = \Delta x$$

$$\Delta x_2 = -\Delta x$$

taking positive displacement as pointing to the right. The force on the mass at any time is therefore:

$$F = -k_1\,\Delta x_1 - (-k_2\,\Delta x_2) = -k_1\,\Delta x + k_2(-\Delta x)$$

$$= -(k_1 + k_2)\,\Delta x$$

$$= -k'\,\Delta x$$

where we let $k' = k_1 + k_2$

Since the frequency of an oscillator having force constant k' is:

$$\nu = \frac{1}{2\pi}\sqrt{\frac{k'}{m}}$$

then: $$\nu = \frac{1}{2\pi}\sqrt{\frac{k_1 + k_2}{m}}$$

Problem 9. Let the force constant k of the spring in the figure be 24 newtons/metre, and let the mass of the body be 4 kgm. The body is initially at rest, and the spring is initially unstretched. Suppose that *a* constant force P of 10 newtons is exerted on the body, and that there is no friction. What will be the speed of the body when it has moved 0.50 metres?

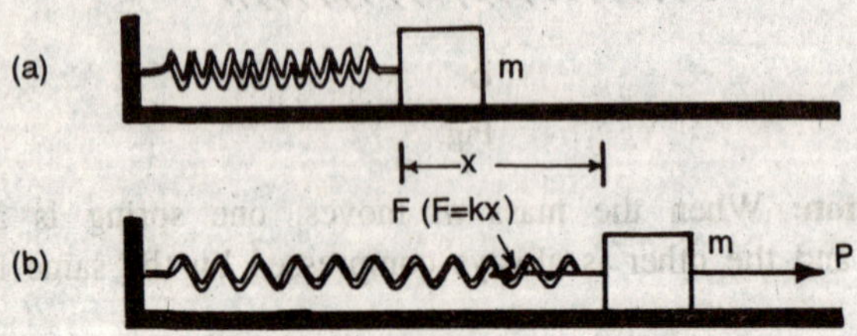

Fig. (a), (b)

Solution: The equations of motion with constant acceleration cannot be used, since the resultant force on the body varies as the spring is stretched. However, the speed can be found using the work-energy theorem. This states that the work done by all the forces acting on a body is equal to the change in kinetic energy of the object. Then, it V_F is the final speed of the block

$$\text{Work} = \int \vec{P}\cdot\vec{dx} + \int \vec{F}\cdot\vec{dx} = \tfrac{1}{2}m\left(V_F^2 - 0^2\right) = \Delta E_k$$

since the initial kinetic energy of the body is zero. P is constant, the angle between $\overrightarrow{P}$ and dx is 0° and the angle between $\overrightarrow{F}$ and $\overrightarrow{dx}$ is 180°. Also the magnitude of the restoring force is F = kx, by Hookes law. Then

$$P\int dx - \int kx\,dx = \tfrac{1}{2}\,mV_F^2$$

$$Px - \tfrac{1}{2}\,kx^2 = \tfrac{1}{2}\,mV_F^2$$

$$10n \times 0.5\,m - \tfrac{1}{2} \times 24\,\frac{n}{m} \times 0.25\,m^2 = \tfrac{1}{2} \times 4kg \times V_F^2$$

or

$$V_F = 1\,\frac{m}{sec}$$

Problem 10. Geologists on a plane are attempting to locate the exact position of the iron reserves in a region by measuring the variation of g in the area, since a large mass of iron will exert an appreciable additional gravitational attraction on a body in the vicinity. They hover above selected spots and observe the movement of a mass suspended from a light spring. If the system has a natural period of 2 s, and the smallest movement of the mass which can be detected is 10^{-6} m, what is the minimum change in g which they can observe?

Solution: At a point at which the acceleration due to gravity is g, the mass, when in equilibrium, has two forces acting on it: the weight mg down and the restoring force $F = kx$ (i.e. Hooke's law) up, and these must be equal. Thus $g = (k/m)x$.

At another point, where local conditions vary, the stretching of the spring will be $x + dx$ if the value of the acceleration due to gravity is $g + dg$ (where dx and dg) are very small changes in x and in g, respectively.

$$\therefore \quad g + dg = \frac{k}{m}(x + dx) \quad \text{or } dg = \frac{k}{m}\,dx.$$

But, if the mass-spring system is allowed to oscillate, its period is given by

$$T = \frac{1}{f} = \frac{2p}{2pf} = \frac{2\pi}{\omega} = 2p\sqrt{\frac{m}{k}}$$

$$\frac{k}{m} = \frac{4\pi^2}{T^2} \text{ and } dg = \frac{4\pi^2}{T^2}\, dx$$

If the smallest value of dx observable is 10^{-6} m, and we are given that T = 2 sec, the smallest value dg detectable is thus

$$dg = \frac{4\pi^2}{4\,s^2} \times 10^{-6}\ \text{m} = 9.87 \times 10^{-6}\ \text{m . s}^{-2}.$$

Problem 11. A pendulum which has a period of 1 s in London, where $g = 32.200$ ft . s^{-2}, is taken to Paris, where it is found to lose 20 s per day. What is the value of g in Paris?

Solution: Since the period of the pendulum is 1 s in London, the number of oscillations it performs per day is

$$\frac{60\ sec}{1\ \text{min}} \times \frac{60\ \text{min}}{1hr} \times \frac{24hr}{1day} = 86{,}400 \text{ oscillations/day}$$

In Paris it looses 20 s per day, i.e., it makes only 86,380 complete oscillations per day.

In all pendulum formulas the period is $T = k/\sqrt{g}$, where k is a constant depending on the shape and possibly the mass of the pendulum. Thus, in London, $T = k/\sqrt{g}$, and in Paris, $T' = k/\sqrt{g'}$. Since T = $1/f$, where f is the number of oscillations in a given time interval.

$$\frac{T}{T'} = \frac{1/f}{1/f'} = \frac{f'}{f} = \frac{86{,}380 \text{ oscillations/day}}{86{,}400 \text{ oscillations/day}} = \sqrt{\frac{g'}{g}}$$

or

$$g' = \left(\frac{8638}{8640}\right)^2 g = \left(\frac{8640-2}{8640}\right)^2 g = \left(1 - \frac{2}{8649}\right)^2 g$$

$$= \left(1 - \frac{4}{8640}\right) g$$

since $(2/8640) << 1$.

Hence, $g' \approx (1 - 4/8640) \times 32.200 \text{ ft} \cdot \text{s}^{-2}$

$= 32.185 \text{ ft} \cdot \text{s}^{-2}$.

Neglecting rotational effects, Paris is slightly farther away from the centre of the earth than London.

Problem 12. What is the acceleration of gravity at a place where the period of a simple pendulum 100 cm long is exactly 2 sec?

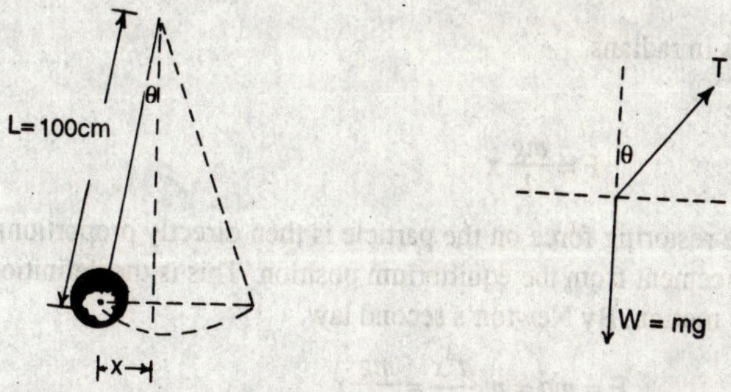

Fig.

Solution: We assume that the pendulum bob are the tension of the string and the weight of the bob (it's mass is *m*). We assume that the displacement of the bob from the equilibrium position $x = 0$ (see figure) is small enough that the bob undergoes only a horizontal motion (i.e., the arc of motion of the bob is approximated by a straight line). The *y* direction of motion is then zero. Therefore there is no net force acting in the *y* direction. Hence

$$T \cos \theta = mg \text{ or } T = \frac{mg}{\cos\theta} \qquad (1)$$

The net restoring force acting in the x direction is

$$F = T \sin \theta$$

$$= mg \frac{\sin\theta}{\cos\theta}$$

$$= mg \tan \theta$$

Here we made use of equation (1). Since the displacement of the bob is very small, then the following approximation can be made.

$$\sin \theta \approx \theta \qquad \text{and} \qquad \cos \theta \approx 1$$

$$F \approx mg\, \theta$$

The restoring force is then directly proportional to the angular displacement θ. Since the horizontal displacement x is approximately equal to the arc legnth of a circle of radius 1, then

$$x = 1\theta$$

where θ is in radians.

Therefore

$$F = \frac{mg}{l} x$$

The restoring force on the particle is then directly proportional to the displacement from the equilibrium position. This is the definition of harmonic motion. By Newton's second law,

$$F = ma = m\frac{d^2x}{dt^3} = \frac{mg}{l}x$$

$$\frac{d^2x}{dt^2} = \frac{g}{l}x$$

The solution of this differential equation, is

$$x = A \sin\sqrt{\frac{g}{l}}\, t + B\cos\sqrt{\frac{g}{l}} = C\sin\left(\sqrt{\frac{g}{1}}t + \phi\right) \quad (1)$$

This can be verified by substitution into the differential equation. The constants A and B (or alternatively c and ϕ) can be found from the initial conditions of the problem (i.e., where the bob was located and its velocity, at $t = 0$). Equation (1) is of the from

$$x = c\sin(\omega t + \phi)$$

Therefore $\quad w = 2\pi f = \sqrt{\frac{l}{g}}$

where f is the frequency of motion. However $f = \frac{1}{T}$ where T is the time taken by the bob to undergo one oscillation. Then

$$\frac{2\pi}{T} = \sqrt{\frac{g}{l}}$$

$$T = 2\pi\sqrt{\frac{l}{g}}$$

$$2^2 = \frac{(2\pi)^2(100)}{g}$$

therefore

$$g = \frac{4 \,.\, \pi^2 \,.\, 100\, cm}{4s^2} = 1000 \text{ cm/sec}^2.$$

Problem 13. One end of a fingernail file is clamped in a vise and the other end is given a to-and-fro vibration. The motion of the free end is approximately S.H.M. If the frequency is 10 vibrations per second and the amplitude is 4 millimeters, what is the velocity when the displacement of the free end is 2 millimeters?

Solution. The problem states that the motion is S.H.M. Therefore, we know that the displacement of the file is

$$x = \text{A} \sin(\omega t + \alpha) \qquad (1)$$

where A is the amplitude and a is a constant. ω is the angular frequency of the vibration. If f is the frequency of the motion

$$\omega = 2\pi f.$$

The velocity of the end of the file is, differentiating (1)

$$v = \text{A}\, \omega \cos(\omega t + \alpha) \qquad (2)$$

We need the velocity when $x = 2$ mm. At this position, using (1)

$$2 \text{ mm} = 4 \text{ mm} \sin(\omega t + \alpha)$$

$$\sin(\omega t + \alpha) = 1/2$$

whence $(\omega t + \alpha) = 30°$.

Hence, using (2),

$$v = \text{A}\, \omega \cos(30°)$$

or $$v = \text{A}\,(2\pi f) \cos(30°)$$

$$v = (4 \text{ mm})\,(6.28)\,(10 \text{ per sec})\,(\sqrt{3}/2)$$

$$v = \left(\frac{40mm}{sec}\right)(6.28)\,(.866) = 218 \text{ mm/sec.}$$

Problem 14. What must be the length of a simple pendulum that will have a period of 1 sec at the surface of the Earth?

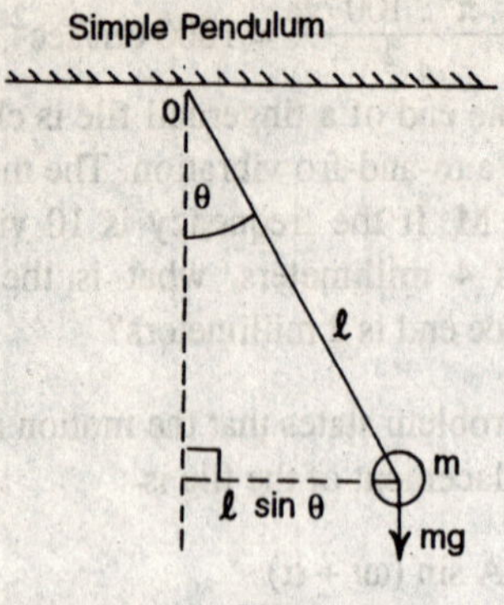

Fig.

Solution: In the simple pendulum figure, the torque about point 0 is

$$\tau = -mgl \sin \theta,$$

where l is the length of the string, and the negative sign is introduced to show that the torque acts to decrease θ.

For θ less than 30°, θ = sin θ is a good approximation thus:

$$\tau = -mgl\,\theta$$

We note that this equation resembles the standard form for simple harmonic oscillators:

$$F = -kx$$

where mgl is analogous to k, and θ to x.

We find the expression for the period T of a simple pendulum by analogy with the equation for that of the harmonic oscillator.

$$T = 2\pi\sqrt{\frac{m}{k}}$$

thus:

$$T = 2\pi\sqrt{\frac{I}{mgl}} = 2\pi\sqrt{\frac{ml^2}{mgl}} = 2\pi\sqrt{\frac{l}{g}}$$

where $I = ml^2$ is the moment of inertia of the mass m about point 0.

From the above equation:

$$l = \frac{T^2 g}{4\pi^2}$$

$$l = \frac{(1\ sec)^2\ (9.8\ m/sec^2)}{4(3.14)^2} = 0.248 \text{ m.}$$

Problem 15. What is the period of a small oscillation of an ideal pendulum of length *l*, if it oscillates in a truck moving in a horizontal direction with acceleration *a*? (see figure a).

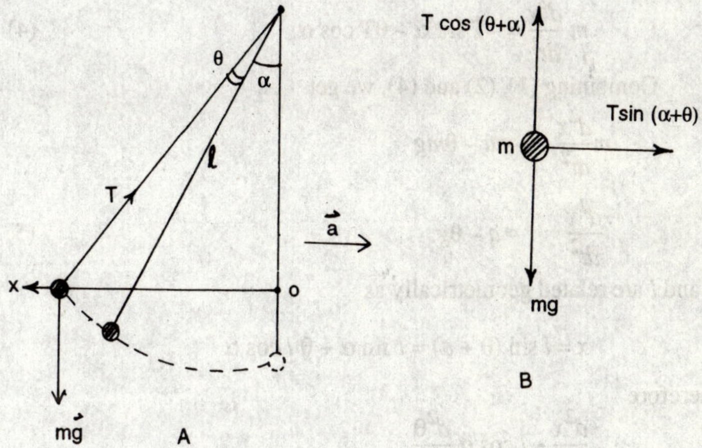

Fig.

Solution: Let the equilibrium position be given by the angle α. In this position, the net force on *m* is along the horizontal axis and equals *ma*. The angle α is determined by the equations.

$$\text{T} \sin \alpha = ma \qquad (1)$$

$$\text{T} \cos \alpha = mg \qquad (2)$$

where *g* is the gravitational acceleration. When the pendulum is displaced by a small amount θ, it will perform a simple harmonic motion around the equilibrium position. The force on *m* along the horizontal x-axis is T sin $(\theta + \alpha)$, as shown in Fig. b. Newton's second law for the motion along this axis is

$$m\frac{d^2x}{dt^2} = -\,\mathrm{T}\sin(\theta + \alpha) \tag{3}$$

where x is the distance from the vertical.

For small θ, we can expand $\sin(\theta + \alpha)$ as

$$\sin(\theta + \alpha) = \sin\alpha + \theta\cos\alpha.$$

Substituting in (3)

$$m\frac{d^2x}{dt^2} = -\mathrm{T}\sin\alpha - \theta\mathrm{T}\cos\alpha. \tag{4}$$

Combining (1), (2) and (4), we get

$$m\frac{d^2x}{dt^2} \quad -ma - \theta mg$$

$$\frac{d^2x}{dt^2} \simeq a - \theta g. \tag{5}$$

θ and l are related geometrically as

$$x = l\sin(\theta + a) = l\sin\alpha + \theta\, l\cos\alpha$$

therefore

$$\frac{d^2x}{dt^2} - l\cos\alpha\,\frac{d^2\theta}{dt^2}$$

Substituting in (5),

$$l\cos\alpha\,\frac{d^2\theta}{dt^2} -\alpha - \theta g$$

$$\frac{d^2\theta}{dt^2} \simeq \frac{g}{l\cos\alpha}\left[\theta + \frac{a}{g}\right] \tag{6}$$

If we make the following substitution in (6)

$$\phi = \theta + \frac{a}{g}$$

we get

$$\frac{d^2\phi}{dt^2} = -\frac{g}{l\cos\alpha}\,\phi.$$

This is the differential equation for a simple harmonic motion. Therefore its solution is

$$\phi = A \sin(\omega t + B)$$

where A and B are the constants of integration and

$$\omega = \sqrt{\frac{g}{l\cos\alpha}}\,.$$

The expression for q is

$$\theta = A \sin(\omega t + B) - \frac{a}{g}.$$

The boundary conditions are $\theta = 0$ at $t = 0$ and $\theta = \theta_{max}$ at $t = \frac{\tau}{2}$ (where τ is the period).

Determine the constants A, B:

(i) $t = 0$: $A \sin B = \frac{a}{g}$

(ii) $t = \frac{\tau}{2}$: $A \sin\left(\frac{\omega t}{2} + B\right) - \frac{a}{g} = \theta_{max}$.

The period of the motion is

$$\tau = \frac{2\pi}{\omega} = 2\pi\sqrt{\frac{l\cos\alpha}{g}}\,.$$

Problem 16. A ball of mass 100 g is attached to the end of string

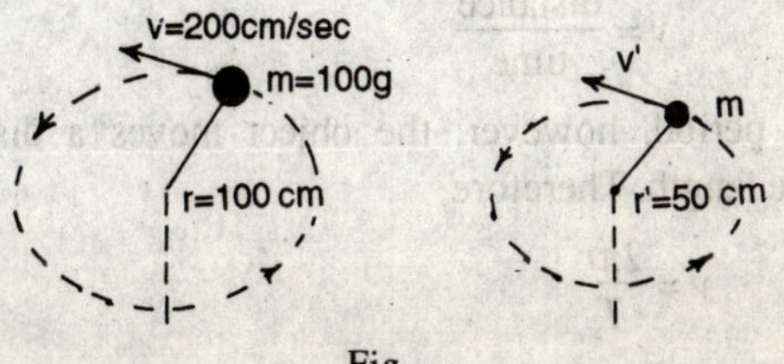

Fig.

and is swung in a circle of radius 100 cm with a constant linear velocity of 200 cm/sec. While the ball is in motion, the string is shortened to 50 cm. What is the change in the velocity and in the period of the motion?

Solution: We define the initial state of the ball to be the circular path of 100 cm radius and the final state to be the path of 50 cm radius. The initial angular momentum of the object is

$$L = mvr$$

$$= (100 \text{ g}) \times (200\ cm/sec) \times (100\ cm)$$

$$= 2 \times 10^6 \text{ g–cm}^2/\text{sec}$$

The initial period is $T = \frac{1}{f}$ where f is the frequency of rotation of the ball in its initial motion. But

$$2\pi f = \omega$$

where ω is the initial angular velocity of the ball. Hence

$$T = \frac{1}{f} = \frac{2\pi}{\omega}$$

Also, $\omega = v/r$ where v is the velocity of the ball and r is its distance from the axis of rotation. Therefore

$$T = \frac{2\pi}{(v/r)} = \frac{2\pi r}{v}$$

$$= \frac{2\pi \times (100cm)}{200\ cm/sec}$$

$$= \pi \text{ sec}$$

An alternate derivation is as follows. The velocity v of the object about the initial circular path is constant. Hence,

$$v = \frac{\text{distance}}{\text{time}}$$

In one period, however, the object moves a distance of one circumference length. Therefore

$$v = \frac{2\pi r}{T}$$

or
$$T = \frac{2\pi r}{v}$$

Shortening the string does not apply any torque to the ball because the applied force lies along the line connecting the ball with the center of rotation. Therefore, the torque is zero. However,

$$\tau = \frac{d\vec{L}}{dt}$$

Then
$$0 = \frac{d\vec{L}}{dt}$$

or
$$\vec{L} = \text{constant.}$$

Hence the final angular momentum, L′, is equal to the initial angular momentum L:

$$L' = mv'\,r' = L$$

Thus, the final velocity is

$$v' = \frac{L}{mr'}$$
$$= \frac{2 \times 10^6\ g\text{–}cm^2/sec}{(100g) \times (50\ cm)}$$

The new period is

$$T' = \frac{2\pi r'}{v},$$
$$= \frac{2\pi \times (50cm)}{400cm/sec}$$
$$= \frac{\pi}{4}\ \text{sec.}$$

Therefore, decreasing the radius by a factor of 2 has increased the linear velocity by the same factor but has decreased the period by a factor of 4.

Problem 17. The figure represents a small body of mass m revolving in a horizontal circle with velocity v of constant magnitude at the end of a cord of length L. As the body swings around its path, the cord sweeps over the surface of a cone. The cord makes an angle θ with the vertical, so the radius of the circle in which the body moves is R = L

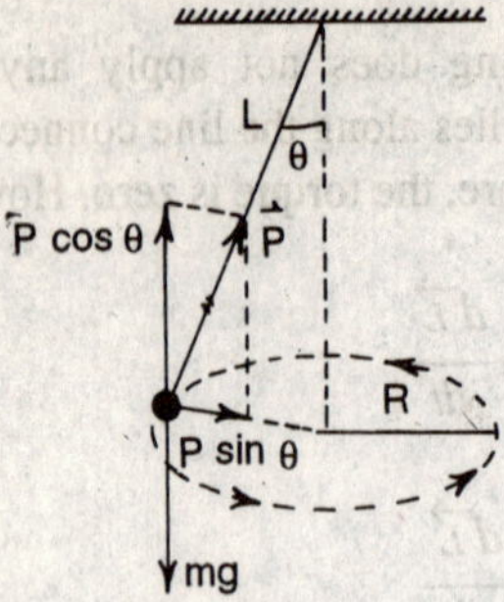

Fig.

sin θ and the magnitude of the velocity v, equals $v = 2\pi R/T = (2\pi L \sin \theta)/T$, where T is the period of revolution of the motion, the time for one complete revolution. Find T.

Solution: The forces exerted on the body when in the position shown are its weight mg and the tension P in the cord. Let P be resolved into a horizontal component P sin θ and a vertical component P cos θ. The body has no vertical acceleration, so the forces P cos θ and mg are equal, and the resultant inward, radial, or centripetal force is the component P sin θ. Then

$$P \sin \theta = m\frac{v^2}{R}, \; P \cos \theta = mg.$$

When the first of these equations is divided by the second, we get

$$\tan \theta = \frac{v^2}{Rg}.$$

This equation indicates how the angle depends on the velocity v. As v increases, tan θ increases and θ increases. The angle never becomes 90°, however, since this requires that $v = -\infty$.

Making use of the relations R = L sin θ and $v = 2\pi L \sin \theta/T$, we can also write

$$\tan \theta = \frac{\sin\theta}{\cos\theta} = \frac{v^2}{Rg} = \frac{4\pi^2 L^2 \sin^2 \theta}{T^2 L \sin \theta \, g}$$

$$\frac{\tan\theta}{\sin\theta} = \frac{1}{\cos} = \frac{4\pi^2 L}{gT^2}$$

$$\cos\theta = \frac{gT^2}{4\pi^2 L}$$

$$T = 2\pi\sqrt{L\cos\theta/g}.$$

This equation is similar in form to the expression for the time of swing of a simple pendulum for which $T = 2\pi\sqrt{L/g}$. Because of this similarity, the present device is called a conical pendulum.

Problem 18. Two springs are joined and connected to a mass m which rides on a frictionless surface (see figure (a)). What is the frequency v of the oscillation that will result if the mass is displaced a small distance x?

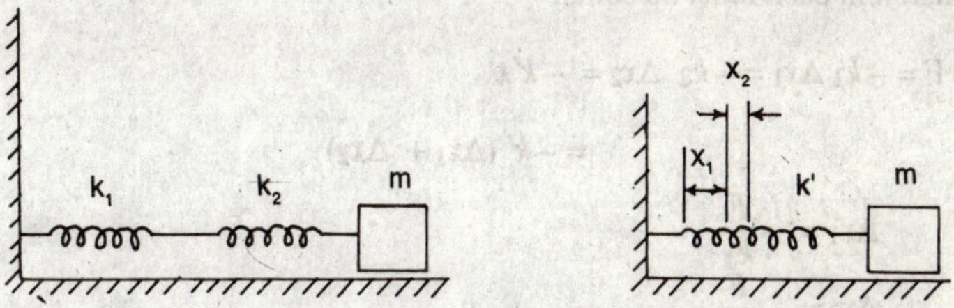

Fig.

Solution: We must show that the configuration of figure (a), is equivalent to that of mass m, connected to a single spring having force constant k'. In figure (b), a mass m is shown connected to such a spring. A segment of the spring is divided into two sections of length x_1 and x_2. The tension in these two sections must be equal, as must be true of the tension in any other arbitrary sections into which the spring may be divided. If this were not the case, the spring would buckle. In general, when the spring is stretched or compressed by a force F, the changes in length Δx_1 and Δx_2 of sections x_1 and x_2 will not be equal since the magnitudes of such distortions will be proportional to the lengths of the segments involved. We can see this by symmetry. If a spring is stretched to twice its usual length, for example, each arbitrary section must stretch to twice its usual length. If we choose the sections to be of unequal length, the amount that each section stretches will be different. Thus:

$$F = -k_1\,\Delta x_1 = -k_2\,\Delta x_2$$

since we know that the force that the spring exerts on the mass equals the tension within each section. The force constants of sections x_1 and x_2 are represented by k_1 and k_2.

It is not unreasonable to assume that each arbitrary section of the spring can be considered to have a different force constant as long as:

$$F = -k_1 \Delta x_1 = -k_2 \Delta x_2 = -k_3 \Delta x_3 = \ldots - k_n \Delta x_n$$

$$= -k'x$$

where $x = \Delta x_1 + \Delta x_2 + \Delta x_3 + \ldots + \Delta x_n$ is the total displacement of the mass.

Thus we see that in the case of the two connected springs above, the tension in each must be equal:

$$F = -k_1 \Delta x_1 = -k_2 \Delta x_2 = -k'x$$

$$= -k'(\Delta x_1 + \Delta x_2)$$

$$\Delta x_1 = -\frac{F}{k_1}$$

$$\Delta x_2 = -\frac{F}{k_2}$$

Thus:

$$k' = -\frac{F}{\Delta x_1 + \Delta x_2} = \frac{-F}{\frac{-F}{k_1} - \frac{F}{k_2}} = \frac{1}{\frac{1}{k_1} + \frac{1}{k_2}}$$

$$= \frac{1}{\frac{k_1 + k_2}{k_1 k_2}} = \frac{k_1 k_2}{k_1 + k_2}$$

Since the frequency of an oscillator having force constant k' is:

$$\nu = \frac{1}{2\pi}\sqrt{\frac{k'}{m}}$$

then:

$$\nu = \frac{1}{2\pi}\sqrt{\frac{k_1 k_2}{(k_1 + k_2)m}}$$

Problem 19. A metal disc is suspended from its centre by a torsion bar, as shown in the figure. Find the period of oscillation for small angular displacements from equilibrium. Torsion bars have the property of exerting a torque which is proportional to the angular displacement and oppositely directed.

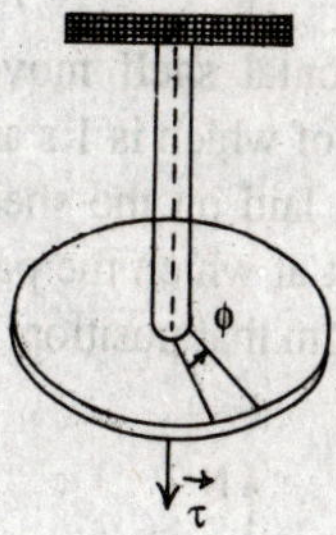

Fig.

Solution: The angular displacement *f* is opposed by a torque

$$\tau = -\kappa\,\phi$$

where κ is the torsion constant, and corresponds to the force constant of a tensile spring. The angular acceleration α and the moment inertia of the disc are related to τ by

$$\tau = I\,\alpha = I\,\frac{d^2\phi}{dt^2}$$

hence $$I\,\frac{d^2\phi}{dt^2} = -\kappa\,\phi\,.$$

The resulting differential equation for ϕ is

$$\frac{d^2\phi}{dt^2} + \Omega^2\,\phi = 0$$

where $\Omega = \sqrt{\frac{\kappa}{I}}$. This equation defines a simple harmonic motion, given by

$$\phi(t) = \phi_0 \sin \Omega t,$$

ϕ_0 is the maximum angular displacement and we assumed that $\phi = 0$ at $t = 0$. The period T is

$$T = \frac{2\pi}{\Omega} = 2\pi \sqrt{\frac{I}{\kappa}}\ .$$

Problem 20. A horizontal shelf moves vertically with simple harmonic motion, the period of which is 1 s and the amplitude of which is 30 cm. A light particle is laid on the shelf when it is at its lowest position. Determine the point at which the particle leaves the shelf and the height to which it rises from that position, g being taken as π^2 m.s^{-2}.

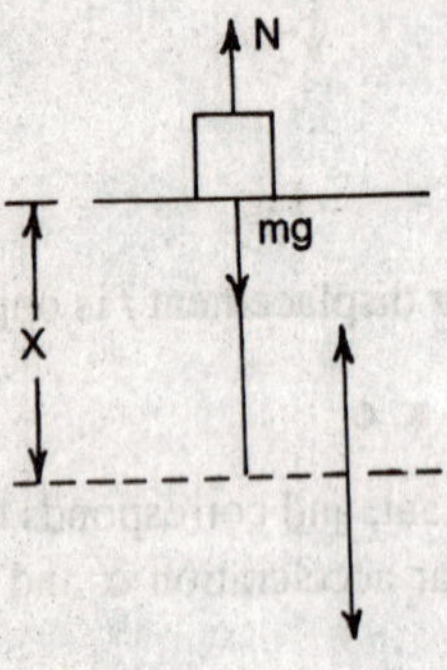

Fig.

Solution: The period of a simple harmonic motion is given by the expression

$$T = \frac{1}{f} = \frac{2\pi}{2\pi f} = \frac{2\pi}{\omega}$$

Therefore, the angular frequency in this case is $\omega = 2\pi/T = 2\pi/1\text{s} = 2\pi$ rad. s^{-1}.

The only forces acting on the particle are its weight mg downward and the normal force $\vec{N}$ exerted by the shelf upwards. (See the figure). At any time, according to Newton's second law, N – mg = ma, where a is the upward acceleration of shelf and particle. If $a = -g$, (i.e. the shelf accelerates downward with magnitude g) N becomes zero and, if a

becomes more negative, the shelf is retarded at a greater rate than the particle; therefore the particle moves away from the shelf. Since the shelf undergoes simple harmonic motion, its displacement may be described by x = A cos (ωt + δ), where A is the amplitude of the oscillation, and d is a phase factor dependent on the initial conditions (i.e. where the shelf is at $t = 0$). The acceleration a $d^2x/dt^2 = -\omega^2$ A cos $(\omega t + \delta) = -\omega^2 x$. Hence, the displacement x at which the particle leaves the shelf is given by

$$-g = -\omega^2 x \qquad \text{or}$$

$$x = \frac{g}{\omega^2} = \frac{\pi^2\ m\ .\ s^{-2}}{4\pi^2\ rad^2\ .\ s^{-2}} = \frac{1}{4}\ \text{m}.$$

$$(\text{for } g = \pi^2\ \text{m} \cdot \text{s}^{-2} \text{ and } \omega = 2\pi\ \text{rad} \cdot \text{sec}^{-1})$$

The particle thus leaves the shelf when it is ¼ m above the mean position. At that point the common velocity of shelf and particle is given by the formula relating to velocity to displacement for simple harmonic motion:

$$v = \pm\ \omega \sqrt{A^2 - x^2}$$

Since the shelf is rising, v is positive and

$$v = 2\pi\ \text{rad}\ .\ \text{s}^{-1} \times \sqrt{(0.3m)^2 - (0.25m)^2}$$

$$= 2p\sqrt{0.0275}\ \text{m} \cdot \text{s}^{-1}.$$

We now have a new problem concerning a particle thrown upward from a platform with an initial speed π. If the platform level is taken as the reference level for measuring potential energy, then the law of conservation of energy requires that at each moment of time that the particle is in motion, the sum of its kinetic and potential energies must remain constant. At the time the particle leaves the platform, its potential energy is zero and

$$E_T = PE + KE = 0 + \tfrac{1}{2} mv^2$$

At its maximum height h, $v = 0$ and

$$E_T = PE + KE = mgh + 0$$

$$\therefore \quad \tfrac{1}{2} mv^2 = mgh \qquad \text{or}$$

$$h = \frac{v^2}{2g} = \frac{4\pi^2 \times 0.0275\ m^2 \ .\ s^{-2}}{2\times \pi^2\ m\ .\ s^{-2}} = 0.055\text{m} = 5.5\text{cm}.$$

Problem 21. A vertical spring has an unstretched length L. When a mass m hangs at rest from its lower end, its length increases to L + l. Find the period of small vertical oscillations of m (figure).

Solution: When the mass is hanging at rest (figure b), the

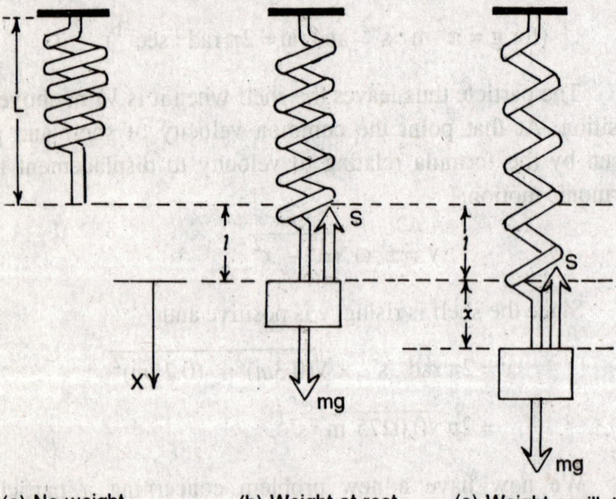

Fig.

extension of the spring is l and the force exerted by it on m is, according to Hooke's law,

$$S = -kl.$$

The positive direction is downward, but S is an upward force and therefore negative. Since the mass has no acceleration, this upward force exerted by the spring must be exactly counterbalanced by the weight, mg and by Newton's Second Law,

$$F_{net} = -kl + mg = 0.$$

Hence,

$$mg = kl. \qquad (1)$$

Suppose that, during the oscillation, the mass is a distance x below its equilibrium position, so that the extension of the spring is $l + x$ (figure c). The force exerted on m by the spring is then

$$S' = -k\,(l + x)$$

$$= -kl - kx.$$

The total force F on m is the sum of S′ and the weight mg.

$$F = S' + mg$$

$$F = -kl - kx + mg.$$

According to equation (1) $-kl$ cancels $+mg$, so

$$F = -kx \qquad (2)$$

where x is the extension of the spring from its equilibrium position (figure c). In order to find the period of small vertical oscillations of m, we must solve the equation of motion (2) for $x\,(t)$. Noting that

$$F = ma = m\frac{d^2x}{dt^2}$$

we may write, from (2)

$$m = \frac{d^2x}{dt^2} = -kx$$

$$\frac{d^2x}{dt^2} + \left(\frac{k}{m}\right)x = 0 \qquad (3)$$

We define

$$\omega_0^2 = \frac{k}{m} \qquad (4)$$

Using (3) and (4),

$$\frac{d^2x}{dt^2} + \omega_0^2 x = 0 \tag{5}$$

This is a linear, second order differential equation for x in terms of the variable t. A valid method of solution is to make an educated guess for x(t), substitute this guess into (5), and see if an identity results. If so, $x(t)$ is the solution of (5). A good guess for $x(t)$ is

$$x(t) = A\cos(at + \phi). \tag{6}$$

where a, A and f are arbitrary constants. Noting that

$$\frac{dx(t)}{dt} = -aA\sin(at + \phi)$$

$$\frac{d^2x(t)}{dt^2} = -a^2A\cos(at + \phi)$$

and substituting these results into (5), we obtain

$$-a^2A\cos(at + f) + \omega_0^2 A\cos(at + \phi) = 0$$

which is an identity if $\omega_0^2 = a^2$. Hence, x(t) is a solution if $\omega_0 = a$ and

$$x(t) = A\cos(\omega_0 t + \phi).$$

To find the period of the motion, note that the cosine function goes through one complete cycle of variation when its argument $(\omega_0 t + \phi)$ has gone through 2π radians. Here, the change in the argument is

$$(\omega_0 t_f + \phi) - (\omega_0 t_0 + \phi)$$

or

$$\omega_0 (t_f - t_0)$$

and this must equal 2π radians, or

$$\omega_0 (t_f - t_0) = 2\pi$$

$$t_f - t_0 = \frac{2\pi}{\omega_0} \tag{7}$$

But $t_f - t_0$ is the time required for cosine to go through one cycle, which is defined as the period of the function (T). Using (7) and (4),

$$T = \frac{2\pi}{\omega_0} = 2\pi\sqrt{\frac{m}{k}} \tag{8}$$

From (1), $\quad mg = kl$

or $\quad k = \frac{mg}{l}$

Inserting this in (8), we obtain

$$T = 2\pi\sqrt{\frac{l}{g}}$$

Problem 22. A mass $m = 2$ kg is hung on a spring of constant $k = 3.92 \times 10^3$ N/m. The natural angular frequency of the system is $\omega_0 = 44.3\ \text{sec}^{-1}$. Now we will force the system to vibrate at different frequencies by an alternating force

$$F = 2\cos\omega t \text{ newtons}$$

That is, the maximum force is two newtons. A static force of two newtons would cause a deflection of about ½ mm to the system. What will be the amplitude of vibration for $\omega = 15\ \text{sec}^{-1}$, and $60\ \text{sec}^{-1}$?

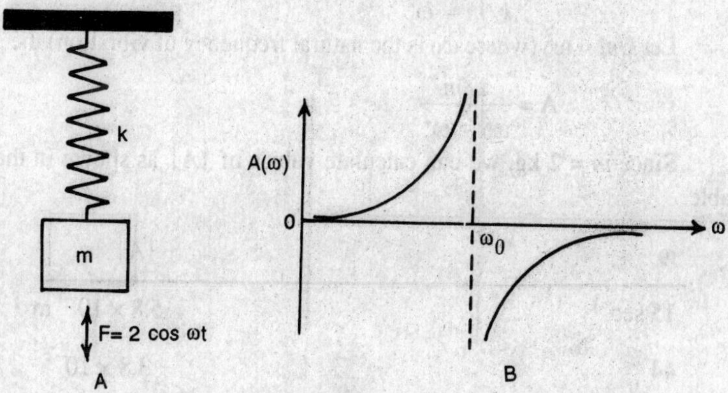

Fig.

Solution: This problem is a case of harmonic motion with a periodic force applied to a vibrating mass. This forces acting on the mass m are the given applied force $F = 2\cos\omega t$, and the restoring force

$F_R = -ky$ due to the spring. y is the vertical displacement of the mass from its rest position. The negative sign indicates that the restoring force acts in such a direction as to oppose a change in vertical displacement. By Newton's second law,

$$-ky + 2\cos\omega t = m\frac{d^2y}{dt^2}$$

or

$$\frac{d^2y}{dt^2} + \frac{k}{m}y = \frac{2}{m}\cos\omega t \tag{1}$$

The following function is a solution to the above differential equation,

$$y(t) = A\cos\omega t \tag{2}$$

where A is the amplitude of vibration. This can be verified by substitution into equation (1):

$$-\omega^2 A\cos\omega t = \frac{k}{m}A\cos\omega t = \frac{2}{m}\cos\omega t$$

Dividing out the common factor, and solving for A, we find

$$A = \frac{2/m}{k/m - \omega^2}$$

Let $k/m = \omega_0^2$ (where ω_0 is the natural frequency of vibration) the

$$A = \frac{2/m}{\omega_0^2 - \omega^2}$$

Since $m = 2$ kg, we can calculate values of $|A|$ as shown in the table.

ω	$\lvert A\rvert$
15 sec^{-1}	5.8×10^{-4} m
44	3.8×10^{-2}
60	6.1×10^{-4}

When the frequency ω of the applied force is near the natural frequency ω_0 the amplitude becomes much larger (see figure b). We can also calculate the average total energy <H> of the system. The total energy

H of the system at any moment is

$$H = \frac{1}{2} mv^2 + \frac{1}{2} ky^3 \qquad (3)$$

where the first term is the kinetic energy of the object and the second term is elastic potential energy of the object. Upon differentiating equation (2), substituting it into equation (3), we obtain

$$H = \frac{1}{2} mA^2 \omega^2 \sin^2 \omega t + \frac{1}{2} kA^2 \cos^2 \omega t$$

But $k = \omega_0^2 m$ (by definition of ω_0^2). Therefore

$$H = \frac{1}{2} mA^2 \omega^2 \sin^2 \omega t + \frac{1}{2} mA^2 \omega_0^2 \cos^2 \omega t$$

The average value of $\cos^2 \omega t$ over 1 period is, by definition

$$\langle \cos^2 \omega t \rangle = \frac{1}{2\pi} \int_0^{2\pi} \cos^2 \omega t \; d(\omega t) = \frac{1}{2}$$

Similarly, $\langle \sin^2 \omega t \rangle = 1/2$

Hence $\langle H \rangle = \frac{1}{4} mA^2 \omega^2 + \frac{1}{4} mA^2 \omega_0^2$

or

$$\langle H \rangle = \frac{1}{4} mA^2 (\omega^2 + \omega_0^2)$$

which shows that the energy of the system becomes much larger as $\omega \rightarrow \omega_0$.

ω	$\langle H \rangle$
15 sec^{-1}	3.7×10^{-4}
44	2.8
60	1.0×10^{-3}

As the frequency of the applied force nears the resonance frequency it is possible to transfer considerably greater energy to the system.

Problem 23. Six students each of mass 80 kg decide to bounce an automobile of mass 1200 kg. They find that if they sit in the auto, the static deflection of the chassis is 1.5 cm. When they press down the bumpers in unison a maximum dynamic deflection of 10 cm is obtained. The auto also oscillates at a frequency of 0.8 sec^{-1}. Estimate the Q of the automobile suspension, and the damping factor ρ of the shock absorbers.

Solution: First we find the spring constant k. The total mass of the students is 6 × 80 = 480 kg; and their weight 480 × 9.8 N = Mg causes a deflection of 1.5×10^{-2} m. According to Hooke's law,

$$W = -kx$$

where x is the extension of the spring of the auto from its equilibrium position, and W is the force (the weight of an object in this case) on the spring. Then

$$k = \frac{w}{x} = \frac{Mg}{x} = \frac{480\ kg \times 9.8\ m/sec^2}{1.5 \times 10^{-2} m} = 3.14 \times 10^5\ \text{N/m}$$

Next we estimate the work done by the students in pushing the car down. Suppose that in pushing down they take all their weight off their feet. Then in each push a force of (480 × 9.8) N travels through a distance of 0.10 m (10 cm). The work done in each oscillation is W = 480 × 9.8 0.10 = 470 joules. The frequency is 0.8 sec^{-1}, so that the power expended per oscillation is found as follows:

$$p = \frac{\text{work/oscillation}}{\text{time for one oscillation}} = \omega \times f$$

$$= 470\ \text{j} \times 0.8\ sec^{-1} = 376\ \text{watt}$$

Here we made use of the relation $f = 1/T$.

The potential energy stored in the suspension is given by

$$H = ½\, kA^2$$

where A is the maximum deflection. Thus

$$H = ½\ 3.14 \times 105\ \text{N/m} \times (0.10\ \text{m})^2$$

$$= 1.57 \times 10^3\ \text{joules}$$

From this we can estimate Q of the system.

The Q of the system is a measure of the amount of energy dissipated in the system, per cycle of operation relative to the energy present in the system. More precisely, we define Q as

$$Q = 2\pi \frac{H}{P}$$

where H is the energy stored in the system, and P is the energy dissipated, all in one cycle

In our case we have

$$Q = 2\pi \frac{1.57 \times 10^3 j}{470 J}$$

$$= 21$$

1.57×10^3 j is the potential energy stored in the automobile springs.

Then we assume that f is the natural frequency (a moderately good assumption) we can calculate ρ using an alternate expression for Q.

Q can also be described in terms of the damping factor ρ. The damping factor is a measure of the time taken for the oscillations to die out. This in turn depends on the amount of energy dissipated by the system per oscillation. Q then, in terms of time, rather than in terms of energy, is given as follows:

$$Q = \frac{\omega_0}{\rho} = \frac{2\pi f}{\rho}$$

From this, we can find ρ. Then

$$\rho = 2\pi f/Q = 2\pi \times 0.8/21 = 0.24 \text{ sec}^{-1}.$$

Problem 24. A particle which is performing simple harmonic motion passes through two points 20 cm apart with the same velocity, taking 1 s to get from one point to the other. It takes a further 2 s to pass through the second point in the opposite direction. What are the period and amplitude of the motion? (Fig. on next page)

Solution: From the equations of simple harmonic motion, the velocity at any displacement x has a value $v = \omega\sqrt{A^2 - x^2}$, where ω is the angular frequency of the motion and A its amplitude. Thus v can be the same at two points only if the displacements of these points are $\pm x$. If the origin of the time scale is taken at the mean position of the

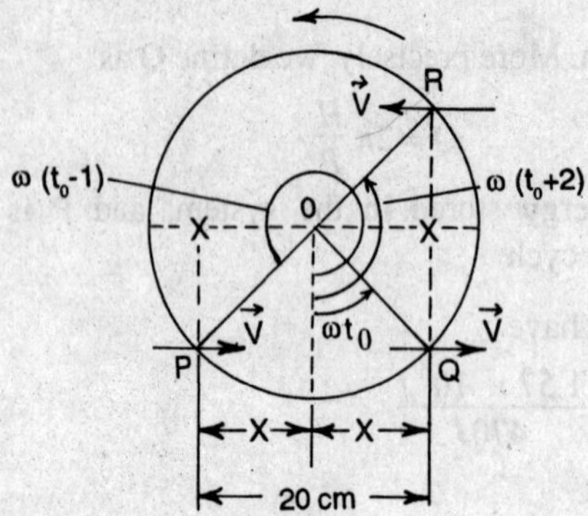

Fig.

motion, as shown in the rotor diagram, and if the second point is reached t_0: thereafter, the displacement at the first point is

$$-x = A \sin \omega (t_0 - 1 \text{ s}) \quad (1)$$

at the second point,

$$+x = A \sin \omega t_0 \quad (2)$$

and at the secomd point on the return journey,

$$+x = A \sin \omega (t_0 + 2\text{s}) \quad (3)$$

Hence, using equations (1) and (2)

$$\sin \omega t_0 = x/A = -\sin \omega (t_0 - 1\text{s})$$

Since

$$-\sin\omega (t_0 - 1s) = \sin\left[-\omega (t_0 - 1s)\right] = \sin \omega t_0$$

$$t_0 = -(t_0 - 1\ s) \quad \text{or} \quad t_0 = ½\ \text{s}.$$

Also, using equations (2) and (3)

$$\sin \omega(t_0 + 2\ s) = x/A = \sin \omega t_0.$$

Since, in general, $\sin \theta_1 = \sin \theta_2$ implies $\theta_2 = \pi - \theta_1$ then, in this particular case

$$\omega\,(t_0 + 2s) = \pi - \omega t_0 \text{ or } \pi/\omega = 2t_0 + 2s = 3s$$

$$\therefore\ \omega = \frac{\pi}{3}\,\text{rad}\cdot\text{s}^{-1}.$$

Also

$$\sin\omega\,(t_0 - 1s) = \sin\left(\frac{\pi}{3}\,rad\,.\,s^{-1} \times -\tfrac{1}{2}\,s\right)$$

$$= -\sin\frac{\pi}{6} = -\tfrac{1}{2}$$

But $\sin\omega\,(t_0 - 1\text{ s}) = -x/\text{A} = -(0.1/\text{A})\text{m}$.

$$\therefore\quad \tfrac{1}{2} = \frac{0.1}{A}\,\text{m} \qquad \text{or} \qquad \text{A} = 0.2\text{ m}.$$

Problem 25. A 10 kg mass resting on a table is attached to a steel wire 5 m long which hangs from a hook in the ceiling. If the table is removed, the wire is momentarily stretched and the mass commences to oscillate up and down. Calculate the frequency of the oscillation if the cross-sectional area of the wire is 1 sq mm and the value of Young's Modulus is taken to be 19×10^{11} dynes/cm^2.

Solution: Young's modulus is defined as

$$\text{Y} = \frac{\dfrac{F}{A}}{\dfrac{\Delta x}{x}}$$

where $\Delta x/x$ is the fractional change in length of the wire, A is its cross-section and F is the force normal to A.

$$\therefore\ \text{F} = \text{Y}\frac{\Delta x}{x}\,\text{A}$$

But F = ma (Newton's second law, where a = acceleration)

$$\therefore\qquad \text{a} = \frac{F}{m} = \left(\frac{Y}{m}\,\frac{A}{x}\right)\Delta x \qquad (1)$$

Since a is proportional to the extension of the wire from its equilibrium length (Δx), the motion is simple harmonic. For this type of motion,

$$a = \omega_0^2\,x \qquad (2)$$

where ω_0 is the angular frequency of the oscillation. Comparing (2) and (1)

$$\omega_0^2 = \frac{YA}{mx}$$

and $$\omega_0 = \sqrt{\frac{YA}{mx}}$$

If f is the frequency of the oscillation

$$\omega_0 = 2\pi f = \sqrt{\frac{YA}{mx}}$$

whence $$f = \frac{1}{2\pi}\sqrt{\frac{YA}{mx}}$$

Using the data supplied

$$f = \frac{1}{2\pi}\sqrt{\frac{(19\times 10^{11}\ dynes/cm^2)\ (1\ mm)^2}{(10\ kg)\ (5m)}}$$

Transforming all quantities to c.g.s values

$$f = \frac{1}{2\pi}\sqrt{\frac{(19\times\ 10^{11}\ dynes/cm^2)\ (10^{-2}\ cm^2)}{(10000\ g)\ (500\ cm)}}$$

$f \approx 10$ vib/s .

Problem 26. Derive the formulas for the period and the frequency of a simple harmonic oscillator.

Solution: A simple harmonic oscillator in one dimension is characterized by a mass which is acted upon by a force, F, proportional to and directed opposite to its displacement x:

$$F = -kx \qquad (1)$$

From Newton's second law:

$$F = ma = m\frac{d^2x}{dt^2} = -kx,$$

$$\frac{d^2x}{dt^2} = -\frac{k}{m}x \qquad (2)$$

From the theory of differential equations, a solution for x is:

$$x(t) = A\cos(\omega t + \delta) \tag{3}$$

where A, ω, and δ are constants. The constants A and δ are arbitrary. However, ω depends upon the physical characteristics of the oscillator, *m* and *k*. We can solve for by inserting the expression for *x* given in equation (3), into equation (2). We must first calculate the second derivative of *x* with respect to *t* :

$$\frac{dx}{dt} = -\omega A \sin(\omega t + \delta), \quad \frac{d^2x}{dt^2} = -\omega^2 A \cos(\omega t + \delta)$$

Then, making the substitution:

$$-\omega^2 A \cos(\omega t + \delta) = -\frac{k}{m} A \cos(\omega t + \delta)$$

Thus, $\omega^2 = \frac{k}{m}, \qquad \omega = \sqrt{\frac{k}{m}}$

Next, we can prove that the oscillator goes through one complete cycle in time T = 2π/ω, by showing that at time *t* + 2π/ω the displacement *x* of the oscillator will be the same as at time *t*:

$$x\left(t + \frac{2\pi}{\omega}\right) = A\cos\left[\omega\left(t + \frac{2\pi}{\omega}\right) + \delta\right]$$

$$= A\cos(\omega t + 2\pi + \delta)$$

$$= A\cos[(\omega t + \delta) + 2\pi]$$

$$= A\cos(\omega t + \delta)$$

The last step is justified from the laws of trigonometry which state that:

$$\cos(y + 2\pi) = \cos y$$

where we let $y = \omega t + \delta$.

The time T it takes the oscillator to complete one cycle is called the period and is represented by T. Thus:

$$T = \frac{2\pi}{\omega} = 2\pi\sqrt{\frac{m}{k}}$$

The frequency of oscillation v, is the number of cycles the oscillator goes through per unit time. This is the reciprocal of the period:

$$\text{cycles/sec} = \frac{1}{\text{sec/cycle}}$$

$$\upsilon = \frac{1}{T} = \frac{\omega}{2\pi} = \frac{1}{2\pi}\sqrt{\frac{k}{m}}$$

Problem 27. In an electric shaver, the blade moves back and forth over a distance of 2 mm, with a frequency of 60 cps. This constitutes a simple harmonic motion with amplitude R = 1 mm. (a) Find the maximum acceleration. (b) Find the maximum velocity. (c) Calculate the velocity and acceleration at point C, a distance x = 0.5 mm from 0, the centre point (see diagram).

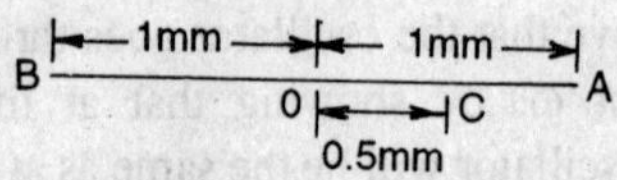

Fig.

Solution: For simple harmonic motion, the force acting on the blade must be proportional to its displacement x, but opposite in direction. The limits of the oscillation are equidistant from the equilibrium position. We can then write

$$F = -kx = ma = m\frac{d^2x}{dt^2} \tag{1}$$

where k is a constant, m is the mass of the blades, and the negative sign indicates that the force acts in the opposite direction to the displacement. Equation (1) can be written

$$\frac{d^2x}{dt^2} = -\frac{k}{m}x \tag{2}$$

which requires a solution where the second derivative of $x(t)$ is proportional to $-x(t)$. The sinusoidal functions have such a property. Assuming that at t = 0 sec, the displacement is at its maximum amplitude R, we can write

$$x = R\cos\omega t \tag{3}$$

where ω is a constant. Differentiating twice,

$$\frac{dx}{dt} = -\,\omega R \sin \omega t \tag{4}$$

$$\frac{d^2x}{dt^2} = -\,\omega^2 R \cos \omega t \tag{5}$$

Substituting equations (3) and (5) into (2), we find

$$-\,\omega^2 R \cos \omega t = -\frac{k}{m} R \cos \omega t$$

$$\omega^2 = \frac{k}{m}.$$

Therefore, if the value for ω^2 is chosen to be k/m, then $x = R \cos \omega t$ is a solution of the equation obtained for the simple harmonic oscillation of the blade. From this solution, we see that ω must be the angular frequency of the motion since the function cos ωt repeats itself after a time $2\pi/\omega$. The velocity and acceleration are dx/dt and d^2x/dt^2, respectively, as found in equations (4) and (5).

(a) Looking at the acceleration function

$$a = -\,\omega^2 R \cos \omega t \tag{6}$$

we note that the maximum value is attained when cos ωt is one or $\omega t = n\pi$ where n is an integer. This occurs when the displacements is a maximum. The angular frequency for the blade is

$$\omega = 2\pi f = (2\pi)\ (60\ sec^{-1}) = 377\ \text{sec}^{-1}$$

Therefore, the maximum acceleration is

$$a_{\max} = \omega^2 R = (377\ \text{sec}^{-1})^2 \times 1\ \text{mm}$$

$$= 1.42 \times 10^5\ \text{mm/sec}^2 = 1.42 \times 10^2\ \text{m/sec}^2$$

(b) To find the maximum velocity, see that

$$v = -\,\omega R \sin \omega t \tag{7}$$

has a maximum whenever sin $\omega t = 1$. This happens whenever the blade is at its equilibrium position at which point $\omega t = n\pi/2$, where n is an integer. The maximum velocity is calculated to be

$$v_{\max} = \omega R = 377\ \text{sec}^{-1} \times 10^{-3}\ \text{m} = 0.377\ \text{m/sec}$$

(c) To express the velocity function in terms of the displacement x, we square equation (7). Then,

$$v^2 = \omega^2 R^2 \sin^2 \omega t = \omega^2 R^2 (1 - \cos \omega t)$$

$$= \omega^2 (R^2 - R^2 \cos^2 \omega t) = \omega^2 (R^2 - x^2)$$

Expressing this in terms of $v_{max} = \omega R$, we write

$$v^2 = \omega^2 R^2 \left(1 - \frac{x^2}{R^2}\right) = v_{max}^2 \left(1 - \frac{x^2}{R^2}\right)$$

At $x = 0.5$ mm

$$v = v_{max} \sqrt{1 - \frac{x^2}{R^2}} = 0.377 \text{ m/sec} \sqrt{1 - \left(\frac{0.5\ mm}{1.0\ mm}\right)}$$

$$v = 0.377 \sqrt{1 - (½)^2} \text{ m/sec}$$

$$= 0.377 \times 0.865 \text{ m/sec} = 0.326 \text{ m/sec}$$

The acceleration function of equation (6) can be expressed as

$$a = -\omega^2 x$$

and at $k = 0.5$ mm = (5) (10^{-4} m), the acceleration is

$$a = \omega^2 x = (377\ sec^{-1})^2 \times (5 \times 10^{-4} m)$$

$$= 71 \text{ m/sec}^2$$

Problem 28. The period of a compound pendulum is 2 s on the earth's surface. What is its period if it is aboard a rocket accelerating upward with an acceleration of 4.3 m $\cdot$ s^{-2}?

Solution: When on the surface the compound pendulum has a period $T = 2\pi \sqrt{I_0/Mgh}$, where M is the mass of the pendulum, I is its moment of inertia about the axis of rotation, h is the distance between the centre of mass and the axis of rotation and g is the acceleration of gravity, as shown in the figure.

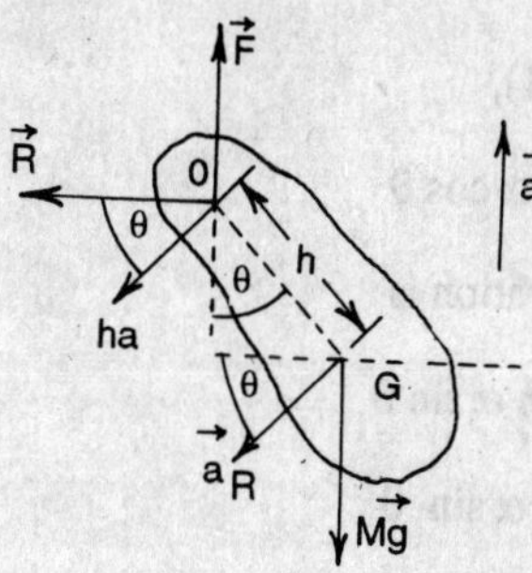

Fig.

Under the acceleration $\vec{a}$ upward, the forces acting on the body when it is displaced through an angle θ are those shown in the figure, the weight Mg downward and the vertical force $\vec{F}$ and horizontal force $\vec{R}$ exerted by the pivot on the pendulum.

Consider the linear accelerations, horizontal and vertical, acting on the body at its centre of mass and the rotational acceleration about the centre of mass. Applying Newton's Law to the horizontal and vertical motions, we get

$$R = Ma_H \tag{1}$$

$$F - Mg = Mav \tag{2}$$

The equation of motion for the rotation is

$$\text{Net torque} = Rh\cos\theta - F\,h\sin\theta\; I_G\,\alpha \tag{3}$$

where I_G is the moment of inertia with respect to the center of mass, and α is the angular acceleration.

The point 0 thus has an acceleration, a_H horizontally, an acceleration av vertically, and a further linear acceleration $a_R = ha$ at right angles to OG, due to the rotation about G. But the point O does not move sideways, only upward with an acceleration a. Thus, there is no net horizontal acceleration

$$a_h + h\,\alpha\cos\theta = 0 \tag{4}$$

Substituting (4) in(1),

$$R = -Mh\,\alpha \cos\theta$$

The upward acceleration is

$$a = a_v - h\,\alpha \sin\theta$$

or $$a_v = a + h\,\alpha \sin\theta \tag{5}$$

Substituting (5) in (2)

$$F = Mg + M\,(a + h\,\alpha \sin\theta)$$

$$= M\,(g + a + h\,\alpha \sin\theta)$$

and finally equation (3) becomes

$$I_G\,\alpha = Rh\cos\theta - Fh\sin\theta$$

$$= Mh^2\,\alpha\cos^2\theta - Mh^2\,\alpha\sin^2\theta - M(g+a)h\sin\theta$$

$$= -Mh^2\,\alpha\,(\cos^2\theta + \sin^2\theta) - M\,(g+a)\,h\sin\theta$$

$$= -Mh^2\,\alpha - M(g+a)\,h\sin\theta$$

or $(I_G + Mh^2) = \alpha = I_0\alpha = M\,(g+a)\,h\sin\theta$

where we used the equation $I_0 = I_G + Mh^2$

But the angle θ is small, and $\sin\theta$ can be replaced by θ,

$$\alpha = \frac{d^2\theta}{dt^2} = -\frac{M(g+a)h}{I_0}\sin\theta = -\frac{M(g+a)h}{iI_0}\,\theta$$

This is the differential equation of a simple harmonic motion, therefore it follows from the theory of simple harmonic motion that the period of oscillation is

$$T' = 2\pi\sqrt{\frac{I_0}{M(g+a)h}}$$

If T is the period of the motion when the pendulum is at rest on the earth

$$\frac{T'}{T} = \frac{2\pi\sqrt{\dfrac{I_0}{M(a+g)h}}}{2\pi\sqrt{\dfrac{I_0}{Mgh}}}$$

$$= \sqrt{\frac{g}{g+a}} = \sqrt{\frac{9.8\ m.s^{-2}}{9.8\ m.s^{-2} + 4.3\ m.s^{-2}}}$$

$$= 0.83$$

Hence T′ = 0.83 T = 0.83 × 2 s = 1.67 s.

Note that the result can be obtained more quickly if the idea of an accelerated frame of reference is applied. An observer in the rocket considers the point of support of the pendulum at rest relative to himself. To explain the observed equilibrium he finds it necessary to postulate a force Ma acting downward on the body in addition to the weight M$\vec{g}$. Hence the pendulum acts as if the weight were M $(\vec{g} + \vec{a})$ insetead of M$\vec{g}$, from which the formula follows immediately.

Problem 29. The simple harmonic motion of the pendulum in figure (a) is slowed as a result of air friction. If the frictional force is proportional to the velocity of the bob, (a) find the displacement of the pendulum as a function of time, (b) calculate the rate of energy

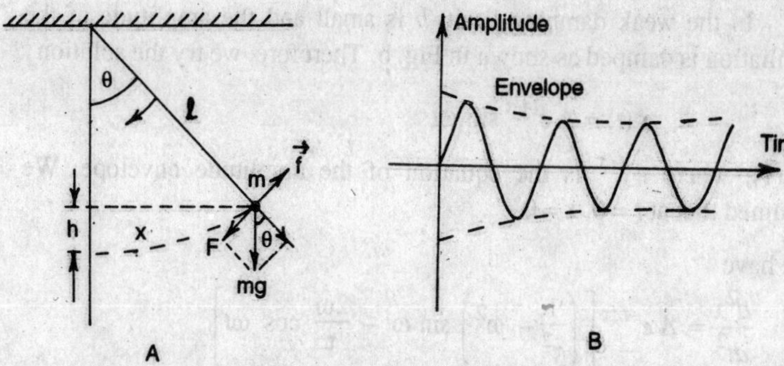

Fig.

dissipation by the damped harmonic motion of the pendulum. Assume that the oscillation frequency of the pendulum is much greater than the rate of damping (the weak damping limit).

Solution: (a) The restoring force on the pendulum for a small angular displacement θ is

$$F = -mg \sin\theta \simeq -mg\,\theta$$

where m and g are the mass of the pendulum and the gravitational acceleration, respectively. The frictional force is

$$f = -bv$$

where b is a proportionality constant, and v is the velocity. The minus sign in the expressions for F and f means that they oppose an increase in θ and v, respectively. The displacement along the arc is

$$x = l\theta$$

The equation of motion for displacement x is

$$m\frac{d^2x}{dt^2} = F + f = -mg\,\theta - bv$$

$$= -mg\frac{x}{l} - b\frac{dx}{dt}$$

which gives the following differential equation for x:

$$\frac{d^2x}{dt^2} + \frac{b}{m}\frac{dx}{dt} + \frac{g}{l}x = 0 \qquad (1)$$

In the weak damping limit, b is small and the amplitude of the oscillation is damped as shown in Fig. b. Therefore, we try the solution

$$x(t) = A\,e^{-t/\tau}\sin\omega t$$

in (1), where $e^{-t/\tau}$ is the equation of the amplitude envelope. We assumed that at $t = 0$, $x = 0$.

We have

$$\frac{d^2x}{dt^2} = A\,e^{-t/\tau}\left[\left(\frac{1}{\tau^2} - \omega^2\right)\sin\omega t - \frac{2\omega}{\tau}\cos\omega t\right]$$

$$\frac{b}{m}\frac{dx}{dt} = A\ e^{-t/\tau}\left[\frac{b\omega}{m}\cos\omega t - \frac{b}{m\tau}\sin\ \omega t\right]$$

$$\frac{g}{l}x = A\,e^{-t/\tau}\frac{g}{l}\ \sin\ \omega t.$$

Substituting in (1), we obtain

$$\mathrm{A}\,e^{-t/\tau}\left\{\left(\frac{1}{\tau^2} - \omega^2 - \frac{b}{m\tau} + \frac{g}{l}\right)\sin\omega t + \left(\frac{b\omega}{m} - \frac{2\omega}{t}\right)\cos\ \omega t\right\} = 0$$

The coefficients of sin ωt and cos ωt must equal to zero separately;

$$\frac{b\omega}{m} - \frac{2\omega}{\tau} = 0$$

$$\frac{1}{\tau^2} - \omega^2 - \frac{b}{m\,\tau} + \frac{g}{l} = 0$$

or

$$\frac{1}{\tau} = \frac{b}{2m} \tag{2}$$

$$\omega^2 = \frac{g}{l} + \frac{1}{\tau^2} - \frac{b}{m\tau} = \frac{g}{l} + \frac{b^2}{4m^2} - \frac{b^2}{2m^2}$$

$$= \frac{g}{l} - \frac{b^2}{4m^2}$$

Hence the angular velocity of the motion is

$$\omega = \sqrt{\omega_0^2 - \frac{b^2}{4m^2}} = \omega_0\sqrt{1 - \left(\frac{b}{2m\omega_0}\right)^2}\ , \tag{3}$$

where $\omega_0^2 = \frac{g}{l}$. For weak damping, $\frac{b}{2m\omega_0} = \frac{1}{\omega_0\,\tau} << 1$, and (3) becomes

$$\omega = \omega_0\left(1 - \frac{b}{m\omega_0}\right) = \omega_0 - \frac{b}{m} \tag{4}$$

(b) The kinetic energy κ is

$$\kappa = ½\ mv^2 = ½\,m\left(\frac{dx}{dt}\right)^2$$

$$= \tfrac{1}{2}\, m \left\{ A\, e^{-t/\tau} \left(\omega \cos \omega t - \frac{1}{\tau} \sin\ \omega t \right) \right\}$$

$$= \tfrac{1}{2}\, m A^2\ e^{-2t/\tau}$$

$$\left(\omega^2 \cos^2\ \omega t + \frac{1}{\tau^2} \sin^2\ \omega t - \frac{2\omega}{\tau}\ \sin\ \omega t\ \cos\ \omega t \right) \qquad (5)$$

Now, let us consider the time average of κ.

For this we substitute the following trignometric identities in (5)

$$2 \sin \omega t \cos \omega t = \sin\ 2\omega t,$$

$$\cos^2\ \omega t = \frac{1 + \cos\ 2\omega t}{2}$$

$$\sin^2\ \omega t = \frac{1\ -\ \cos\ 2\omega t}{2}$$

$$\kappa = \tfrac{1}{2} m\ A^2\ e^{-2t/\tau}$$

$$\left(\tfrac{1}{2}\ \omega^2 + \frac{1}{2\tau^2} + \tfrac{1}{2}\ \omega^2\ \cos\ 2\omega t - \frac{1}{2\tau^2}\ \sin\ 2\omega t - \frac{\omega}{\tau}\ \sin\ 2\omega t \right)$$

The time average of a pure sine wave is zero, i.e.

$$< \cos 2\ \omega t > = < \sin 2\ \omega t > = 0$$

Therefore, only constant terms in the expression for κ survive when we take its time average

$$< \kappa > = \tfrac{1}{2} m A^2\ e^{-2t/\tau} \left(\omega^2 + \frac{1}{\tau^2} \right).$$

We see that the average kinetic energy decays exponentially. The potential energy U is

$$U = mgh = mgl\,(\,1 - \cos\theta) - \tfrac{1}{2} mgl\,\theta^2$$

$$= \tfrac{1}{2}\ mgl \left(\frac{x}{l} \right)^2 = \tfrac{1}{2}\ \frac{mg}{l}\ x^2$$

$$= \tfrac{1}{2}\ \frac{mg}{l} A^2\, e^{-2t/\tau}\ \sin^2\ \omega t$$

$$= \tfrac{1}{4}\ \frac{mg}{l} A^2\ e^{-2t/\tau}\ [1 - \cos 2\omega t]$$

The time average of ∪ is

$$<\cup> = \tfrac{1}{4}\,\frac{mg}{l}\,A^2\,e^{-2t/\tau}$$

Therefore, the time average of total energy is

$$<E> = <\cup> + <\kappa> = \tfrac{1}{4}\,A^2\,e^{-2t/\tau}\left(\frac{mg}{l} + m\omega^2 + \frac{m}{\tau^2}\right)$$

$$= \tfrac{1}{2}\,A^2\,e^{-2t/\tau}\left(\frac{mg}{l} = \frac{mg}{l} - \frac{b^2}{4m} + \frac{b^2}{4m}\right)$$

$$= \tfrac{1}{2}\,\omega_0^2\,A^2\,e^{-2t/\tau}.$$

The rate of change of energy with time has a negative sign since the energy decreases. Therefore, the rate of dissipation, P, is positive since it is the rate of decrease (negative rate of increases) of the energy

$$<P> = -\frac{d}{dt}<E> = \frac{\omega_0^2\,A^2}{\tau}\,e^{-2t/\tau}$$

$$= \frac{2<E>}{\tau}.$$

Problem 30. At a carnival, the people who go on a certain ride sit in chairs around the rim of a horizontal circular platform which is oscillating rapidly with angular simple harmonic motion about a vertical axis through its centre. The period of the motion is 2s and the amplitude .2 rad.

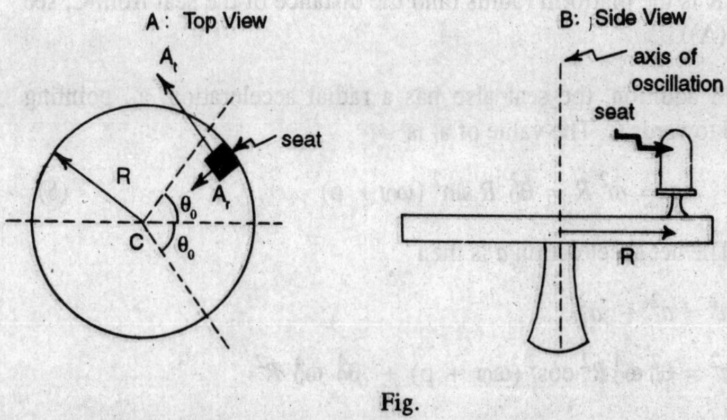

Fig.

One of the chairs becomes unbolted and just starts to slip when the angular displacement is a maximum. Calculate the coefficient of static friction between chair and platform. (The rim is 12 ft. from the center).

Solution: In order to find the coefficient of static friction, we must use Newton's Second Law to relate the acceleration of the seat to the net force acting on it. Now, the platform and seat are oscillating about an axis through C (see figure (A)) with a maximum angular displacement θ_0. Since the motion is simple harmonic, we may write

$$\theta = \theta_0 \cos(\omega_0 t + \rho) \qquad (1)$$

which gives us the angular displacement of a point on the disc (and, hence, the seat) as a function of time. Note that ρ is an arbitrary constant, which is determined by the way the angular motion begins, and ω_0 is the angular frequency of the motion. Differentiating (1) twice

$$\omega = \frac{d\theta}{dt} = -\theta_0\ \omega_0 \sin(\omega_0 t + \rho) \qquad (2)$$

$$\alpha = \frac{d\omega}{dt} = -\ \theta_0\ \omega_0^2 \cos(\omega_0 t + \rho) \qquad (3)$$

Here, ω and α are the angular velocity and acceleration of a point on the disc. Since the seat has an angular acceleration α, it also has a tangential acceleration a_t such that

$$a_t = aR = -\theta_0\ \omega_0{}^2 R \cos(\omega_0 t + \rho) \qquad (4)$$

where R is the platform radius (and the distance of the seat from C, see figure (A)).

In addition, the seat also has a radial acceleration, a_r, pointing inward towards C. The value of a_r is

$$a_r = \omega^2 R = \theta_0^2\ R \sin^2(\omega_0 t + \rho) \qquad (5)$$

The net acceleration a is then

$$a^2 = a_t^2 + a_r^2,$$

$$a^2 = \theta_0^2\, \omega_0^4 R^2 \cos^2(\omega_0 t + \rho) + \theta_0^4\ \omega_0^4\ R^2$$

$$\sin^4(\omega_0 t + \rho)$$

$$a^2 = \theta_0^2\ \omega_0^4 R^2 \left[\cos^2(\omega_0 t + \rho) + \theta_0^2 \sin^4\ \omega_0 t + \rho\right]$$

or $$a = \theta_0\ \omega_0^2 R\ \left(\cos^2(\omega_0 t + \rho) + \theta_0^2 \sin^4(\omega_0 t + \rho)\right)^{1/2} \qquad (6)$$

If we assume the chair is unbolted, there are 2 sets of forces acting on the chair. In the vertical direction (perpendicular to the disc) 2 forces act the normal force N of the disc on the chair, and its weight, *mg* (*m* is the chair's mass). Since the chair stays on the platform, there is no acceleration in the vertical direction, and N and *mg* balance:

$$\mathrm{N} = \mathrm{mg} \qquad (7)$$

In the plane of the disc, the only force acting is the force of static friction, f_s. By Newton's Second Law, this must equal the product of the chair's mass and acceleration. Using (6)

$$f_s = m\,\theta_0\ \omega_0^2\ R\left(\cos^2(w_0 t + \rho) + \theta_0^2\ \sin^4(\omega_0 t + \rho)\right)^{1/2}$$

But, $f_s \le u_s\ N$

Hence, $f_s \le u_s\ mg$

Combining the first, and last equation for f_s,

$$m\,\theta_0\ \omega_0^2 R\ \left(\cos^2(\omega_0 t + \rho) + \theta_0^2 \sin^4(\omega_0 t + \rho)\right)^{1/2} \le u_s\, mg$$

Solving for u_s

$$u_s \ge \frac{\theta_0\ \omega_0^2 R}{g}\left(\cos^2\ (\omega_0 t + \rho) + \theta_0^2 \sin^4\ (\omega_0 t + \rho)\right)^{1/2} \qquad (8)$$

This relation holds as long as the seat is stationary. Once the seat begins to slip, the maximum force of static friction is encountered, and (8) becomes a strict equality:

$$u_s = \frac{\theta_0\ \omega_0^2 R}{g}\ \left(\cos^2(\omega_0 t + \rho) + \theta_0^2 \sin^4(\omega_0 t + \rho)\right)^{1/2} \qquad (9)$$

Note that slippage occurs at the maximum value of θ, or θ_0. For this value of θ, (1) yields

$$\theta_0 = \theta_0 \cos(\omega_0 t + \alpha)$$

whence $\cos(\omega_0 t + \alpha) = 1$

As a result, $\sin(\omega_0 t + \alpha) = 0$

Using these facts in (9)

$$u_s = \frac{\theta_0\ \omega_0^2 R}{g}\ (1+0)^{1/2} = \frac{\theta_0\ \omega_0^2\ R}{g} \qquad (10)$$

Note that we don't know ω_0, but we do know T, the period of the oscillation. By definition

$$\omega_0 = 2\pi f = \frac{2\pi}{T}$$

where f is the oscillation frequency. Using this in (10)

$$u_s = \frac{4\pi^2\ \theta_0 R}{T^2 g}$$

Substituting the given data

$$u_s = \frac{(4)\ (9.89)\ (.2\ rad)\ (12\ ft)}{(2s)^2\ (32\ ft/s^2)}$$

$$u_s = .74$$

We could have done as easier (but less general) analysis by noting that, when $\theta = \theta_0$, the a_r term is zero, for the chair is instantaneously at rest.

Problem 31. Calculate the average over time of the kinetic and potential energy of a harmonic oscillator.

Solution: Assume that the harmonic oscillator we are dealing with is a mass on a spring. The position of the mass is then given by

$$x = A\ \sin(\omega_0 t + \varphi)$$

where A is the amplitude of the motion of the mass, and φ is a phase angle defining the initial position of the mass (at $t = 0$).

The kinetic energy is then

$$K = \tfrac{1}{2}\, Mx^2 = \tfrac{1}{2} M\left[\omega_0 A \cos\,(\omega_0 t + \varphi)\right]^2,$$

The time average of the kinetic energy over one period T of the motion is defined as

$$\langle K\rangle = \frac{\int_0^T K\,dt}{T}$$

But ω_0 is the angular frequency and

$$\omega_0 = 2\pi f$$

where f is the frequency of motion of the mass. Since $f = 1/T$

$$\omega_0 = 2\pi/T$$

and

$$T = 2\pi/\omega_0$$

Then

$$\langle K\rangle = \tfrac{1}{2}\, M\omega_0^2 A^2 \; \frac{\int_0^{2\pi/\omega_0} \cos^2(\omega_0 t + \varphi)\, dt}{2\pi/\omega_0}$$

Let $\quad y = \omega_0 t + \varphi \qquad (1)$

and

$$dy = \omega_0\, dt$$

since φ is a constant. We then have

$$\langle K\rangle = \tfrac{1}{2}\, M\omega_0^2 A^2 \; \frac{\int_{\varphi}^{\varphi + 2\pi} \cos^2 y\,(dy/\omega_0)}{2\pi/\omega_0} \qquad (2)$$

where the limits are obtained by noting that, from (1), $y = \varphi$ at $t - 0$, and $y = 2\pi + \varphi$ at $t = 2\pi/\omega_0$. Rewriting (2)

$$\langle K\rangle = \tfrac{1}{2}\,\frac{M\omega_0^2 A^2}{2\pi}\left(\int_{\varphi}^{\varphi + 2\pi} \cos^2 y\; dy\right)$$

But $\cos^2 y = \frac{1}{2}(1 + \cos 2y)$ and

$$\langle K\rangle = \frac{M\omega_0^2 A^2}{4\pi}\left(\int_{\varphi}^{\varphi+2\pi} \frac{1}{2}(1+\cos 2y)\,dy\right)$$

$$\langle K\rangle = \frac{M\omega_0^2 A^2}{8\pi}\left[\int_{\varphi}^{\varphi+2\pi} dy + \int_{\varphi}^{\varphi+2\pi} \cos 2y\,dy\right]$$

$$\langle K\rangle = \frac{M\omega_0^2 A^2}{8\pi}\left[\int_{\varphi}^{\varphi+2\pi} dy + \frac{1}{2}\int_{\varphi}^{\varphi+2\pi} \cos 2y\,(2dy)\right]$$

$$\langle K\rangle = \frac{M\omega_0^2 A^2}{8\pi}\,[2\pi + \tfrac{1}{2}[\sin(2\varphi + 4\pi) - \sin(2\varphi)]]$$

Since $\sin(2\varphi + 4\pi) = \sin 2\varphi$

$$\langle K\rangle = \frac{M\omega_0^2 A^2}{4}$$

The potential energy is, with $x = A\sin(\omega_0 t + \varphi)$

$$U = \tfrac{1}{2}Cx^2 = \tfrac{1}{2}CA^2\sin^2(\omega_0 t + \varphi)$$

where C is the spring constant. Now,

$$\langle U\rangle = \frac{\int_0^T U\,dt}{T} = \frac{\int_0^{2\pi/\omega} 0\ \frac{1}{2}CA^2\sin^2(\omega_0 t + \varphi)\,dt}{2\pi/\omega_0}$$

Using (1)

$$\langle U\rangle = \tfrac{1}{2}CA^2\ \frac{\int_{\varphi}^{\varphi+2\pi} \sin^2 y\,(dy/\omega_0)}{2\pi/\omega_0}$$

$$\langle U\rangle = \frac{CA^2}{4\pi}\int_{\varphi}^{\varphi+2\pi} \sin^2 y\,dy$$

Because $\sin^2 y = \frac{1}{2}(1 - \cos 2y)$

$$\langle U\rangle = \frac{CA^2}{4\pi}\left[\int_{\varphi}^{\varphi+2\pi} \tfrac{1}{2}(1-\cos 2y)\,dy\right]$$

$$<U> = \frac{CA^2}{8\pi}\left[\int_{\varphi}^{\varphi+2\pi} dy - \int_{\varphi}^{\varphi+2\pi} \cos\, 2y\, dy\right]$$

$$<U> = \frac{CA^2}{8\pi}\left[\int_{\varphi}^{\varphi+2\pi} dy - \tfrac{1}{2}\int_{\varphi}^{\varphi+2\pi} \cos 2y\ (2dy)\right]$$

$$<U> = \frac{CA^2}{8\pi}\ [2\pi - \tfrac{1}{2}\,[\sin(2\varphi + 4\pi) - \sin(2\varphi)]]$$

Since $\sin(2\varphi + 4\pi) = \sin(2\varphi)$

$$<U> = \frac{CA^2}{4}$$

But $\omega_0^2 = C/M$, by definition of the frequency of the simple harmonic oscillator. Therefore

$$<U> = \frac{M\omega_0^2 A^2}{4}$$

Thus $<U> = <K>$ and the total energy of the harmonic oscillator is

$$E = <K> + <U> = \tfrac{1}{2} M\omega_0^2 A^2$$

Note that $E = <E>$ because the total energy is a constant of the motion.

The equality of the average kinetic energy and potential energies is a special property of the harmonic oscillator. This property does not hold in general for an harmonic oscillators. It does hold for weakly damped oscillators.

Problem 32. If a tunnel were drilled through the earth along one of its diameters and if a stone were dropped into it from one end, how

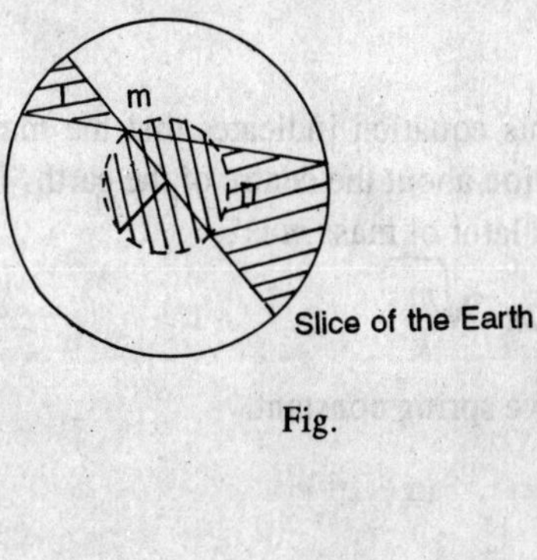

Fig.

long would it be before the stone returned? Compare the answer with the period of an earth satellite in an orbit of minimum radius and comment on the two values. Assume the earth to be of uniform density, and make use of the information that a body inside the earth at a distance r from the centre has a gravitational force acting on it due only to the portion of the earth of radius r.

Solution: Consider the regions I and II of the earth constructed out of the conical sections with pivot m (see the figure). Region I contains a smaller mass than region II. It is, however, closer to m than region II. Since the gravitational force due to a mass is directly proportional to the mass, and inversely proportional to the square of distance, then it appears that regions I and II provide equal and opposite forces on m, due to the trade off between mass and distance. A detailed mathematical analysis verifies that regions I and II do neutralize each other and have no resultant effect on m. This is true for any such regions constructed by the above procedure, and the net effect of the portion of the earth outside the sphere of radius r is zero. At any distance r from the centre the stone is acted on by a force of magnitude

$$F = -\frac{GM'm}{r^2} = -\frac{Gm}{r^2} \times \frac{4}{3}\pi\rho r^3 = -\frac{4}{3}\pi\rho Gmr,$$

where m is the mass of the stone, ρ the density of the earth, and M′ the mass of that portion of the earth of radius r. We also used the fact that $M' = \rho V = \rho(4/3\,\pi r^3)$. The negative sign indicates that the force acts in a direction that opposes an increase in separation of the two bodies (its attractive). Therefore, by Newton's second law,

$$F = ma = -\left(\frac{4}{3}\pi\rho GM\right) r \quad \text{or} \quad F = -kr$$

where $k = 4/3\,\pi\rho Gm$.

The form of this equation indicates that the mass is undergoing simple harmonic motion about the centre of the earth. The period of any simple harmonic oscillator of mass m is

$$T = 2\pi\sqrt{\frac{m}{k}}$$

where k is the effective spring constant.

Therefore, in this case,

$$T = \frac{2\pi}{\sqrt{\frac{4}{3}\pi\rho G}}$$

But at the surface of the earth the gravitational force of attraction on the mass m (i.e. its weight) is

$$W = mg = \frac{GMm}{R^2} = \frac{4}{3}\pi\rho GmR,$$

by Newton's Law of Universal Gravitation.

R is the radius of the earth and M is now the total mass of the earth. Thus

$$\frac{g}{R} = \frac{4}{3}\pi\rho G$$

Hence $T = 2\pi\sqrt{\frac{R}{g}} = 2\pi\sqrt{\frac{6370\times 10^3 m}{9.8\ m.s^{-2}}}$

$$= 5061.6\ s = 84.8\ min.$$

The time period of an earth satellite in a circular orbit is $T' = 2\pi d^{3/2}/R\sqrt{g}$, where d is the distance of the satellite from the centre of the earth. If the height of the satellite above the earth's surface is negligible in comparison with the radius of the earth, then to a first approximation $d = R$, and $T' = 2\pi\sqrt{R/g}$, the same period as that of the stone undergoing simple harmonic motion along the tunnel. This is not surprising, since any simple harmonic motion may be considered the projection on a diameter of the motion of a point in a circle with constant speed. Thus we should expect the period of an earth satellite near the earth's surface to be the same as the period of a particle undergoing simple harmonic motion along a diameter of the circle.

Problem 33. Consider a pendulum which is oscillating with an amplitude so large that we may not neglect the θ^3 term in the expansion of sin θ, as we do for a harmonic oscillator. What is the effect on the motion of the pendulum of the term in θ^3? This is an elementary example of an anharmonic oscillator. Anharmonic or nonlinear problems are usually difficult to solve exactly (except by computers), but approximate solutions are often adequate to give us a good ideas of what is happening.

Solution: The expansion of sin θ (for small θ) to terms of order θ^3, commonly expressed as 'expansion to $0(\theta^3)$,'' is

$$\sin\theta = \theta - \frac{1}{6}\theta^3 + \ldots,$$

so that the equation of motion of a simple oscillator becomes, to this order,

$$\frac{d^2\theta}{dt^2} + \omega_0^2\theta - \frac{\omega_0^2}{6}\theta^3 = 0, \qquad (1)$$

where ω_0^2 denotes the quantity g/L. This is the equation of motion of an anharmonic oscillator.

We see if we can find an approximate solution to (1) of the form

$$\theta = \theta_0 \sin \omega t + \varepsilon\theta_0 \sin 3\omega t, \qquad (2)$$

It is now evident that (2) can only be an approximate solution at best. It remains for us to determine e, and also ω; while ω must reduce to ω_0 at small amplitudes, it may differ at large amplitudes. For simplicity we suppose that $\theta = 0$ at $t = 0$.

An approximate solution of this type to a differential equation is called a perturbation solution, because one term in the differential equation perturbs the motion which would occur without that term. As you have seen, we arrived at the form of (2) by guided guesswork. It is easy enough to try several guesses and to reject the ones which do not work.

We have from (2)

$$\theta = -\omega^2\theta_0 \sin \omega t - 9\omega^2\varepsilon\theta_0 \sin 3\omega t;$$

$$\theta^3 = \theta_0^3(\sin^3\omega t + 3\varepsilon\sin^2\omega t \sin 3\omega t + \ldots)$$

where we have discarded the terms of order ε^2 and ε^3 because of our assumption that we can find a solution with $\varepsilon << 1$. Then the terms of (1) become

$$\omega_0^2\,\theta = \frac{\omega_0^2}{6}\theta^3 - \ddot{\theta}$$

$$\omega_0^2\,\theta_0 \sin \omega_0\tau + \varepsilon\,\omega_0^2\,\theta_0 \sin 3\omega t = \frac{\omega^2}{6}\,\theta_0^3 \times$$

$$(\sin^3 \omega t + 3\varepsilon \sin^2 \omega t \sin 3\omega t) +$$

$$\omega^2\,\theta_0 \sin \omega t + 9\,\omega^2\,\varepsilon\,\theta_0 \sin 3\omega t.$$

The coefficient of the term $\sin^2 \omega t \sin 3\omega t$ in the previous equation is small by $0(\varepsilon)$ or by $0(\theta_0^2)$, compared with the other terms in the equation. Since ε and θ_0 are small quantities, we neglect this term.

Using the trigonometric identity

$$\sin^3 \omega t = \frac{3 \sin \omega t - \sin 3\omega t}{4}$$

we get $\varepsilon w^2\,\theta_0 \sin \omega t + \varepsilon\,\omega_0^2\,\theta_0 \sin 3\omega t = \left(\frac{1}{8}\,\omega_0^2\,\theta_0 + \omega^2\,\theta_0\right) \times$

$$\sin \omega t + \left(9\,\omega^2\,\varepsilon\,\theta_0 - \frac{1}{24}\,\omega_0^2\,\theta_0^3\right) \sin 3\omega t \qquad (3)$$

In order for this equation to hold, the corresponding coefficients of $\sin \omega t$ and $\sin 3\omega t$ on each side of the equation must be equal

$$\omega_0^2 = \frac{\omega_0^2\,\theta_0^2}{8} + \omega^2 \qquad (4)$$

$$\varepsilon\omega^2 = -\frac{1}{24}\,\omega_0^2\,\theta_0^2 + 9\varepsilon\omega^2 \qquad (5)$$

From (4), $\omega^2 = \omega_0^2\left(1 - \frac{1}{8}\,\theta_0^2\right)$

or

$$\omega = \omega_0\left(1 - \frac{1}{8}\,\theta_0^2\right)^{1/2} = \omega_0\left(1 - \frac{1}{16}\,\theta_0^2\right)$$

using the binomial equation for the square root. We see that as $\theta_0 \to 0$, $\omega \to \omega_0$. The frequency shift Δ_ω is

$$\Delta\omega = \omega - \omega_0 = \frac{\omega_0}{16}\theta_0^2$$

In (5), we make the substitution $\omega^2 = \omega_0^2$ and obtain an expression for ε.

$$\varepsilon\,\omega_0^2 = \frac{1}{24}\omega_0^2\,\theta_0^2 + 9\varepsilon\omega_0^2$$

$$\varepsilon = \frac{\theta_0^2}{8 \times 24} = \frac{1}{16}\theta_0^2$$

We think of e as giving the fractional admixture of the sin $3\omega t$ term in a solution for θ dominated by the sin ωt term.

Why did we not include in (2) a term in sin $2\omega t$? Try for yourself a solution of the form

$$\theta = \theta_0 \sin\omega t + \eta\theta_0 \sin 2\omega t,$$

and see what happens. You will find $\eta = 0$. The pendulum generates chiefly third harmonics, i.e., terms in sin $3\omega t$, and not second harmonics. The situations would be different for a device for which the equation of motion included a term in θ^2.

What is the frequency of the pendulum at large amplitudes? There is no single frequency in the motion. We have seen that the most important term (the largest component) is in sin ωt, and we say that ω is the fundamental frequency of the pendulum. To our approximation ω is given by (4). The term in sin $3\omega t$ is called the third harmonic of the fundamental frequency. Our argument following (2) suggests that an infinite number of harmonics are present in the exact motion, but that most of these are very small. The amplitude in (2) of the fundamental component of the motion is θ_0; the amplitude of the third harmonic component is $\varepsilon\theta_0$.

Problem 34. Suppose that the 2 atoms of a stable molecule oscillate along the line connecting their centres. Treating the system as an harmonic oscillator, calculate the vibration frequency. (Neglect any rotation of the system).

Solution: If we treat the given system as an harmonic oscillator,

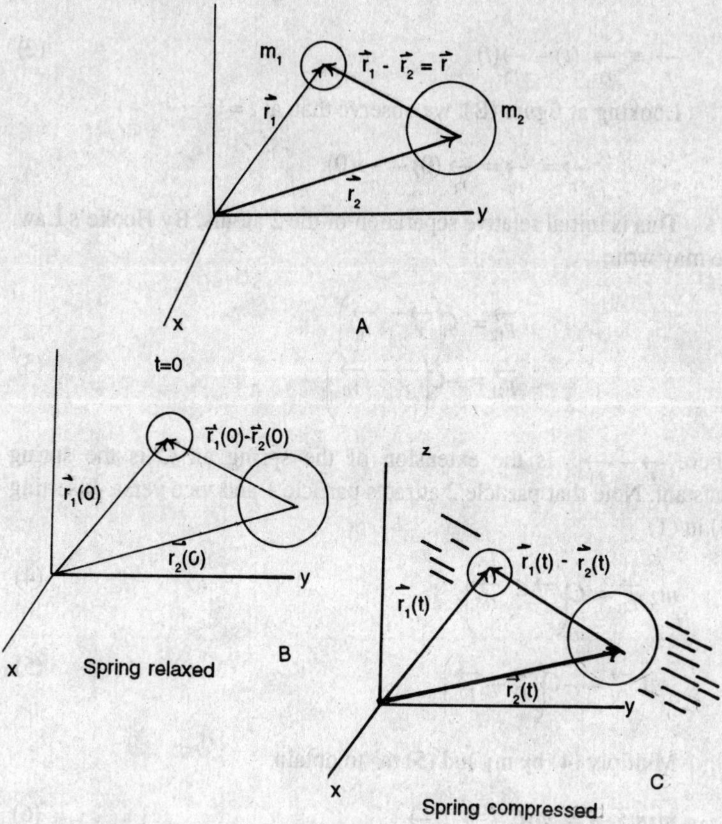

Fig.

we may assume that the 2 atoms are coupled by a tiny spring connecting their centres. Then, using Newton's Second Law to relate the force on each atom to its acceleration, we obtain

$$\overrightarrow{F_{12}} = m_2 \overrightarrow{r_2}(t)$$

$$\overrightarrow{F_{21}} = m_2 \overrightarrow{r_2}(t) \qquad (1)$$

where $\overrightarrow{F_{12}}$ is the force acting on atom 2 due to atom 1, and $\overrightarrow{F_{21}}$ is the force on atom 1 due to atom 2. Also, $\overrightarrow{r_2}$ (t) and $\overrightarrow{r_1}$ (t) are the intial positions of particles 2 and 1, respectively. (See figure (A)). Now, we may define the relative distance between the 2 atoms as

$$\vec{r} \equiv \vec{r_1}(t) - \vec{r_2}(t) \tag{2}$$

Looking at figure (B), we observe that, at $t = 0$,

$$\vec{r} = \vec{r_0} = \vec{r_1}(0) - \vec{r_2}(0).$$

This is initial relative separation of the 2 atoms. By Hooke's Law, we may write

$$\begin{aligned} \vec{F_{12}} &= C\left(\vec{r} - \vec{r_0}\right) \\ \vec{F_{21}} &= -C\left(\vec{r} - \vec{r_0}\right) \end{aligned} \tag{3}$$

where $\vec{r} - \vec{r_1}$ is the extension of the spring an C is the spring constant. Note that particle 2 attracts particle 1 and vice versa. Inserting (3) in (1)

$$m_2 \vec{r_2} = C\left(\vec{r} - \vec{r_0}\right) \tag{4}$$

$$m_1 \vec{r_1} = -C\left(\vec{r} - \vec{r_0}\right) \tag{5}$$

Multiply (4) by m_1 and (5) m_2 to obtain

$$m_1 m_2 \vec{r_2} = m_1 C\left(\vec{r} - \vec{r_0}\right) \tag{6}$$

$$m_1 m_2 \vec{r_1} = -m_2 C\left(\vec{r} - \vec{r_0}\right) \tag{7}$$

Substracting (6) from (7)

$$m_1 m_2 \vec{r_1} - m_1 m_2 \vec{r_2} = m_2 C\left(\vec{r} - \vec{r_0}\right) - m_1 C\left(\vec{r} - \vec{r_0}\right)$$

or

$$m_1 m_2\left(\vec{r_1} - \vec{r_2}\right) = C\left(\vec{r_1} - \vec{r_2}\right)(m_1 + m_2)$$

From (2)

$$\vec{r_1} - \vec{r_2} = \vec{r}$$

and $$\frac{m_1 m_2}{m_1 + m_2}\ddot{\vec{r}} = -C\left(\vec{r} - \vec{r_0}\right)$$

Now m_1m_2/m_1+m_2 is defined as the reduced mass u and

$$u\ddot{\vec{r}} = -C\left(\vec{r} - \vec{r_0}\right)$$

Finally $$\ddot{\vec{r}} + \frac{C}{u}\left(\vec{r} - \vec{r_0}\right) = 0 \qquad (8)$$

Redefining $\vec{r} - \vec{r_0}$ as $\vec{s}$, we note that

$$\ddot{\vec{S}} = \ddot{\vec{r}} - \ddot{\vec{r_0}} = \ddot{\vec{r}}$$

since $\vec{r_0}$ = constant. Equation (8) now becomes

$$\ddot{\vec{s}} + \left(\frac{C}{u}\right)\vec{s} = 0$$

which is the equation of motion of a simple harmonic oscillating of angular frequency

$$\omega_0 = \sqrt{\frac{C}{u}} \qquad (9)$$

It is known from spectroscopic measurements that the fundamental vibrational frequencies of the molecules HF and HCL are

ω_0 (HF) = 7.55×10^{14} rad/sec;

ω_0 (HC1) = 5.47×10^{14} rad/sec;

Let us use these data to compare the force constants C_{HF} and C_{HC1}. The reduced mass of HF is, in atomic mass units,

$$\frac{1}{u_{HF}} = \frac{1}{m_H} + \frac{1}{m_F} = \frac{1}{1\ amu} + \frac{1}{19\ amu} = \frac{20}{19}\ \text{amu}$$

$u_{HF} \approx .950$ amu

$$\frac{1}{u_{HCL}} = \frac{1}{m_H} + \frac{1}{m_{Cl}} = \frac{1}{1\ amu} + \frac{1}{35\ amu} = \frac{36}{35}\ \text{amu}$$

Here we have used the atomic mass of the most abundant isotype of chlorine, Cl^{35}.) Notice that the reduced masses are quite close to each other in value. This is because the hydrogen, being lightest, does most of the oscillating.

Now from (9) we have for the ratio of the force constants:

$$\frac{C_{HF}}{C_{HCl}} = \frac{\left(\mu\,\omega_0^2\right)_{HF}}{\left(\mu\,\omega_0^2\right)_{HCl}} = \frac{54.0\times 10^{28}\ amu/s^2}{29.0\times 10^{28}\ amu/s^2} = 1.86,$$

while for an individual force constant

$$C_{HF} = \left(54\times 10^{28}\ amu/s^2\right)\left(1.66\times 10^{-24}\ g/amu\right) \approx 9\times 10^5\ \text{dyne/cm}$$

Here we have inserted the factor which converts the mass from atomic mass units to gram.

Is this vlaue of C reasonable? Suppose we stretch the molecule (which is about 1A or 1×10^{-8} cm in length) by 0.5 A. The work needed to do thiis would probably be nearly enough to break up the molecule into separate atoms of H and F. By using the formula for the potential energy of a compressed (or extended) spring, the work needed to stretch the molecule 0.5 A should be of the order of magnitude

$$\tfrac{1}{2}C\left(r - r_0\right)^2 \approx \tfrac{1}{2}\left(9\times 10^5\right)\left(0.5\times 10^{-8}\right) \approx 1\times 10^{-11}\ erg$$

This is not unreasonable for an energy of decomposition into separate atoms. Therefore, the calculated value of C is reasonable.

2

Wave Motion

Travelling & Standing Waves

Problem 1. A wave is represented by the equation $y = 0.20 \sin 0.40\pi (x - 60\, t)$, where all distances are measured in centimeters and time in seconds. Find: (1) the amplitude, (b) the wavelength, (c) the speed, and (d) the frequency of the wave, (e) what is the displacement at $x = 5.5$ cm and $t = 0.020$ sec?

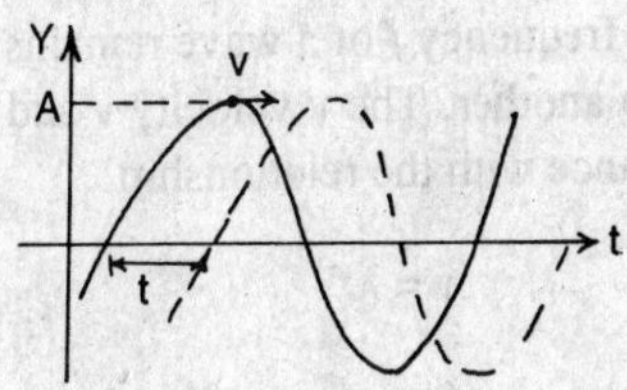

Fig.

Solution: The displacement y of the medium due to wave motion at a position x and at a time t is (see the figure)

$$y = A \sin \frac{2\pi}{\lambda} (x - vt)$$

where A is the amplitude, λ is the wavelength, and v is the velocity with which the wave is travelling along the x-axis. If we compare this equation with the expression given in the question, we see that

(a) $A = 0.20$ cm

(b) $\frac{2\pi}{\lambda} = 0.40\pi$

$$\lambda = \frac{2}{0.40} \text{ cm} = 5.0 \text{ cm}$$

(c) $v = 60$ cm/sec

(d) $$f = \frac{v}{\lambda} = \frac{60\ cm/sec}{5.0cm} = 12/\text{sec}$$

(e) $$y = (0.20\text{ cm}) \sin [0.40\pi\ (5.5 - 60 \times 0.020)]$$

$$= (0.20\text{ cm}) \sin [0.40\pi\ (5.5 - 1.2)]$$

$$= (0.20\text{ cm}) \sin (0.40 \times 4.3\pi)$$

$$= (0.20\text{ cm}) \sin 1.72\pi$$

$$= (0.20\text{ cm})\ (-0.77) = -0.15\text{ cm}.$$

Problem 2. A compressional wave of frequency 250/sec is set up in an iron rod and passes from the rod into air. The speed of the wave is 1.6×10^4 ft/sec in iron and 1.1×10 ft/sec in air. Find the wavelength in each material.

Solution: The frequency f of a wave remains constant as it passes from one medium to another. The v velocity v and wavelength λ of the wave vary is accordance with the relationship

$$v = \lambda f$$

In iron

$$\lambda = \frac{v}{f} = \frac{1.6 \times 10^4\ ft/sec}{250/sec} = 64\text{ ft}$$

In air

$$\lambda = \frac{v}{f} = \frac{1.1 \times 10^3\ ft/sec}{250/sec} = 4.4\text{ ft.}$$

Problem 3. If the frequency of an oscillating source on the surface of a pool of water in 3 Hz, what is the speed of the wave if the wavelength is observed to be 0.5 m?

Solution: An example of an oscillating wave is a sinusoidal wave. Three important properties of an oscillating wave are its velocity of propagation, frequency, and wavelength. Its frequency f is the number of cycles per unit time at which any point oscillates or, expressed in another way, the number of waves that pass a given point per unit time. The wavelength λ is the distance between two adjacent crests of the wave. For a relation between these quantities, note that the time t required for the wave to make one oscillation is $1/f$. During this time, the wave moves a distance $d = \lambda$. From

$$d = vt$$

we have

$$\lambda = v \cdot \frac{1}{f} = \frac{v}{f}$$

or

$$v = f\lambda$$

Substituting the known values,

$$v = (3 \text{ Hz})\,(5 \times 10^{-1} \text{ m}) = 1.5 \text{ m/sec}$$

(Note: $1 \text{ Hz} = \text{sec}^{-1}$)

Problem 4. Find the speed of a compressional wave in an iron rod whose specific gravity is 7.7 and whose Young's modulus is 27.5×10^6 1b/in^2.

Solution: A wave in an iron rod will move with a speed

$$v = \sqrt{\frac{E}{\rho}}$$

where E is Young's modulus, and ρ is the specific weight of iron.

$$E = 27.5 \times 10^6 \text{ 1b/in}^2 = 27.5 \times 10^6 \times \frac{1b}{in^2} \times \frac{144\ in^2}{ft^2}$$

where we' ve used the fact that

$$1 \text{ 1bn/in}^2 = 144 \text{ 1b/ft}^2$$

$$\rho = \frac{D}{g} = \frac{7.7 \times 62.4\ \ 1b/ft^3}{32\ ft/sec^2} = 15.0 \text{ slugs/ft}^3$$

Hence,

$$v = \sqrt{\frac{27.5 \times 10^6 \times 144\ 1b/ft^2}{15.0\ slugs/ft^3}}$$

$= 1.6 \times 10^4$ ft/sec.

Problem 5. The speed of a certain compressional wave in air at standard temperature and pressure is 330 m/sec. A point source of frequency 600/sec radiates energy uniformly in all directions at the rate of 5.00 watts. What is the intensity of the wave at a distance of 20.0 m from the source? What is the amplitude of the wave there?

Solution: At any concentric spherical surface the energy from a point source is spread over an area $4\pi r^2$. At a distance of $r = 20.0$ m, the intensity I is

$$I = \frac{E}{tA} = \frac{P}{A} = \frac{5.00\ watts}{4\pi \times (20.0\ m)^2}$$

$$= 0.99 \times 10^{-3}\ \text{watt/m}^2$$

The rate of transfer of energy depends on the square of the wave amplitude and the square of the wave frequency for all types of waves, with I the average intensity and the density of air

$\rho = 1.29$ gm/litre $= 1.29$ kg/m^3, we have,

$$A^2 = \frac{I}{2\pi^2 v\rho f^2}$$

$$= \frac{0.99 \times 10^{-3}\ watt/m^2}{2\pi^2\ (330\ m/sec)\ (1.29\ kg/m^2)\ (300/sec)^2}$$

$$= 1.32 \times 10^{-12}\ \text{m}^2$$

$$A = 1.15 \times 10^{-6}\ \text{m} = 1.15 \times 10^{-4}\ \text{cm}$$

Problem 6. Standing waves are produced by the superposition of two waves of the form

$$y_1 = 15 \sin (3\pi t - 5x) \quad (1)$$

$$y_2 = 15 \sin (3\pi t + 5x) \quad (2)$$

Find the amplitude of motion at $x = 21$.

Solution: We note that the two waves are of the form

$$y = A \sin(\omega t \pm kx)$$

where y is the displacement of the wave at position x and time t.

A is the amplitude of the wave (i.e., the maximum displacement), ω is the angular frequency ($= 2\pi f$) and k is the angular wavelegnth $\left(= \frac{2p}{\lambda}\right)$ of the wave. Both waves have the same characteristics (i.e., the same amplitude, frequency and wavelength). They differ only in that they travel in opposite directions. The negative sign in wave (1) indicates that it is travelling to the right. Wave (2) is travelling to the left. The resultant wave produced by the superposition of these two waves can be found as follows.

We use the relationships

$$\sin(\alpha + \beta) = \sin\alpha \cos\beta + \cos\alpha \sin\beta$$

$$\sin(\alpha - \beta) = \sin\alpha \cos\beta - \cos\alpha \sin\beta$$

Therefore

$$\sin(\alpha + \beta) + \sin(\alpha - \beta) = 2 \sin\alpha \cos\beta \qquad (3)$$

Thus, comparing (3) with (2) and (1),

$$y = y_1 + y_2 = 2(15) \sin(3\pi) t \cos(5) x$$

$$= 30 \sin 3\pi t \cos 5x$$

This wave pattern is called a standing wave (as opposed to a travelling wave). The wave remains in one location; or alternatively, the energy associated with the wave is not transferred from one location to another. With

$$x = 21, \quad 5x = 105 \text{ radians}$$

$$= 33.4\pi \text{ radians.}$$

Now $\cos 33.4\pi = \cos(.4\pi - 33\pi) = \cos .4\pi = \cos 72^\circ = 0.309$.

Thus , $x = 21$,

$y = 30 \sin 3\pi t \cos 5x$

$y = (3)(.309) . \sin 3\pi t$

$y = 9.27 \sin 3\pi t$

The amplitude of this wave is the maximum value of y. This maximum value is 9.27.

Problem 7. A string 4.0 m long has a mass of 3.0 gm. One end of the string is fastened to a stop, and the other end hangs over a pulley with a 2.0 kg mass attached. What is the speed of a transverse wave in this string?

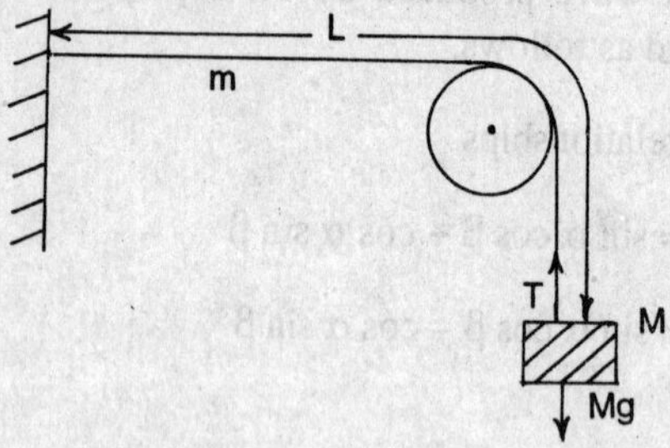

Fig.

Solution: The string is strecthed by a force (the tension), which is inequilibrium with the weight of the mass:

$$T = Mg = 2.0 \text{ kg} \times 9.8 \text{ m/sec}^2 = 19.6 \text{ nt.}$$

The linear density μ (mass per unit length) of the string is

$$\mu = \frac{m}{1} - 0.0030 \text{ kg}/4.0 \text{ m} = 7.5 \times 10^{-4} \text{ kg/m}$$

A transverse wave can be set up in the string by moving the weight at one end up and down. The speed v of this wave is given by

$$v = \sqrt{\frac{T}{\mu}} = \sqrt{\frac{19.6 \ kg\text{–}m/sec^2}{7.5 \times 10^{-4} \ kg/m}}$$

$$= \sqrt{2.6 \times 10^4 \ m^2/sec^2}$$

$$= 160 \text{ m/sec.}$$

Problem 8. An electrical transmission line is strung across a valley between two utility towers. The distance between the towers is 200 m, and the mass of the line is 40 kg. A lineman on one tower strikes the line and a wave travels to the other tower, is reflected, and is detected by the lineman when it returns. If the elapsed time measured by the lineman is 10 s, what is the tension force in the wire?

Solution: The distance d travelled by the wave pulse is $d = 2 \times 200$ m $= 400$ m. The speed of the wave is computed using

$$v = \frac{d}{t} = \frac{400m}{10s} = 40 \text{ m/s}$$

For the stretched line, the tension F is given by

$$F = \mu v^2$$

where μ is the inertia characteristic of the wire and is equal to the mass per unit length m/L. Therefore,

$$F = \frac{m}{L} v^2 = \frac{mv^2}{L} = \frac{(40\ kg)\ (40\ m/s)^2}{200m} = 3.2 \times 10^2 \text{ N.}$$

Problem 9. Two sources separated by 10 m vibrate according to the equations $y_1 = 0.03 \sin \pi t$ and $y_2 = 0.01 \sin \pi t$. They send out simple waves of velocity 1.5 m/sec. What is the equation of motion of a particle 6 m from the first source and 4 m from the second?

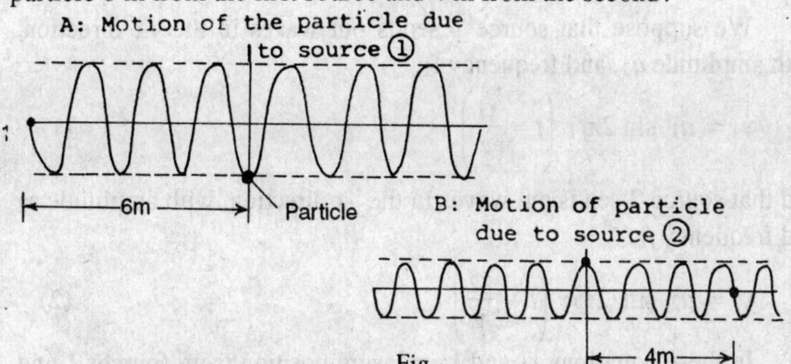

Fig.

Solution: The equation of a wave travelling along the x–axis is

$$y = A \sin (\omega t \pm \kappa x)$$

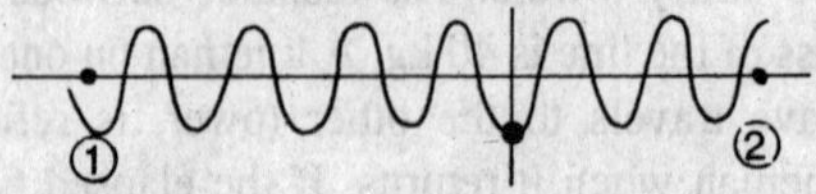

Fig.

where ω is the angular frequency of the wave A is its amplitude and K is its wave number. The factor ±K accounts for the direction of travel of the wave; $-\kappa x$ indicates that the wave travels towards increasing values of x, and vice-versa. Note that this equation assumes that, at $x = 0, t = 0$, the y displacement is zero. Since $\omega = 2\pi f$, where f is the frequency of the wave, and $k = 2\pi/\lambda$, where λ is its wavelength, we obtain,

$$y = A \sin(\omega t \pm Kx)$$

$$y = A \sin\left(2\pi ft \pm \frac{2\pi x}{\lambda}\right)$$

But $\lambda = \frac{y}{f}$ where v is the wavespeed. Therefore,

$$y = A \sin 2\pi\left(ft \pm \frac{fx}{v}\right)$$

$$y = A \sin 2\pi f\left(t \pm \frac{x}{v}\right)$$

We suppose that source 1 sends out waves in the $+x$ direction, with amplitude a_1, and frequency f_1.

$$y_1 = a_1 \sin 2\pi f_1\left(t - \frac{x_1}{v}\right) \tag{1}$$

and that source 2 sends out waves in the $-x$ direction, with amplitude a_2 and frequency f_2.

$$y_2 = a_2 \sin 2\pi f_2\left(t + \frac{x_2}{v}\right) \tag{2}$$

In these equations x_1 and x_2 measure position from sources 1 and 2 as origin, respectively. Comparing equations (1) and (2) with the equations given in the question, we find that,

$$2\pi f_1 = \pi \text{ and } 2\pi f_2 = \pi$$

or

$$f_1 = f_2 = 1/2$$

Furthermore,

$$v = 1.5 \text{ m/sec}$$

$$a_1 = 0.03 \text{ m} \qquad a_2 = 0.01 \text{ m}$$

$$x_1 = 6\text{m} \qquad x_2 = -4 \text{ m}$$

Thus,

$$y_1 = (.03 \text{ m}) \sin \pi (t - 4)$$

$$y_1 = (0.3 \text{ m}) [\sin \pi t \cos 4\pi - \cos \pi t \sin 4\pi]$$

$$y_1 = (.03 \text{ m}) \sin \pi t$$

$$y_2 = (.01 \text{ m}) \sin \pi (t - 8/3)$$

$$y_2 = (.01 \text{ m}) [\sin \pi t \cos 8\pi/3 - \cos \pi t \sin 8\pi/3]$$

$$y_2 = (.01 \text{ m}) [\sin \pi t (-1/2) - \cos \pi t (\sqrt{3}/2)]$$

$$y_2 = -.005 \text{ m} \sin \pi t - .00866 \cos \pi t$$

The resultant wave motion of the particle is,

$$y = y_1 + y_2$$

$$= (.03 \text{ m}) \sin \pi t - (.005 \text{ m}) \sin \pi t - (.00866 \text{ m}) \cos \pi t$$

$$= (.02 \text{ m}) \sin \pi t - .00866 \cos \pi t \qquad (3)$$

We will write this in the form,

$$y = \text{A} \sin (\pi t + \phi)$$

$$= \text{A} \sin \pi t \cos \phi + A \cos \pi t \sin \phi \qquad (4)$$

Comparing (4) with (3),

$$\text{A} \cos \phi = .025 \text{ m}$$

$$A \sin \phi = -.00866 \text{ m}$$

$$A^2 (\sin^2 \phi + \cos^2 \phi) = A^2 = (.025 \text{ m})^2 + (.00866 \text{ m})^2$$

$$A = .0264 \text{ m}$$

and,

$$\tan \phi = \frac{-.00866}{.025} = -.346$$

$$-\tan \phi = .346$$

But,

$$-\tan \phi = \tan(-\phi) \text{ and}$$

$$\tan(-\phi) = .346$$

$$-\phi = \tan^{-1}(.346)$$

$$-\phi = 19.1^\circ$$

ELECTROMAGNETIC WAVES, POLARIZATION

Problem 10. What is the speed of an electromagnetic wave in empty space?

Solution: The speed of an electromagnetic wave in a substance can be determined by using:

$$v = \frac{1}{\sqrt{\mu\varepsilon}}$$

where μ is the permeability and e the permittivity of the substance.

$$\mu_0 = 4\mu \times 10^{-7} \text{ nt/amp}^2$$

$$\varepsilon_0 = 8.55 \times 10^{-12} \text{ coul}^2/\text{nt} \cdot \text{m}^2$$

$$v = \frac{1}{\sqrt{\mu_0 \varepsilon_0}}$$

$$= \frac{1}{\sqrt{4\pi \times 10^{-7} \times 8.85 \times 10^{-12} \; coul^2/amp^2 - m^2}}$$

$$= 3.00 \times 10^8 \text{ m/sec.}$$

Problem 11. H. Hertz produced radio waves whose wavelength was about 3 m. What was the frequency of the oscillating electric charges responsible for this electromagnetic radiation?

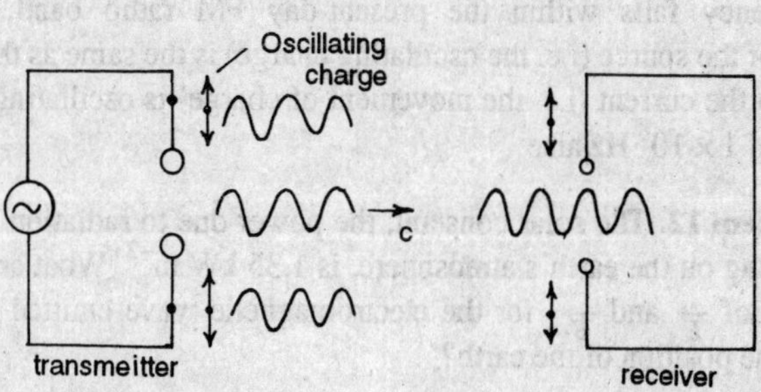

Fig.

Solution: Hertz produced radio waves by use of an induction coil. This alternating current generator was connected in series with a metal loop containing a gap in it (see the figure). A second loop (the receiver) with a similar gap was placed a few feet away from the first loop (the transmitter). The induction coil created an arc discharge across the gap of the transmitting circuit. A similar discharge appeared across the gap of the receiving circuit. Apparently, the disturbance at the first circuit was transmitted to the second circuit. This disturbance was shown to propagate with the speed of light *c*.

The effect of the AC generator is to cause the electric charge within the wire in the transmitting circuit to oscillate about an equilibrium position. An oscillating charge undergoes acceleration and emits electromagnetic radiation with a frequency equal to the oscillating frequency of the charge. The electromagnetic wave disturbance is transmitted through space. When it reaches the receiving circuit, it causes the charge in the wire to oscillate. An AC current is then created. An arc discharge is then developed across the gap due to this current.

If the wavelength of the wave is given, the frequency of the wave can be found by use of the following relation.

$$c = f\lambda$$

Since the value of the wavelength, $\lambda = 3$ m, and the speed of light, $c = 3 \times 10^8$ m/s, are known, the frequency is then

$$f = \frac{c}{\lambda} = \frac{3 \times 10^8\ m/s}{3m} = 1 \times 10^8\ \text{Hz}$$

This frequency falls within the present-day FM radio band. The frequency of the source (i.e. the oscillating charge) is the same as that of the wave so the current (i.e. the movement of charge) is oscillating at a frequency of 1×10^8 Hz also.

Problem 12. The solar constant, the power due to radiation from the sun falling on the earth's atmosphere, is 1.35 kW m^{-2}. What are the magnitudes of $\vec{E}$ and $\vec{B}$ for the electromagnetic wave emitted from the sun at the position of the earth?

Solution: Starting with the electromagnetic waves at the earth, it is possible to determine $\vec{E}$ and $\vec{B}$ by two methods. (a) The Poynting vector

$$\vec{S} \frac{1}{\mu_0} \vec{E} \times \vec{B}$$

gives the energy flow across any section of the field per unit area per unit time.

Here, $\vec{E}$ and $\vec{B}$ are the instantaneous electric field and magnetic induction, respectively, at a point of space, and μ_0 is the permeability of free space. If we approximate the sun as a point source of light, then we realize that it radiates electromagnetic waves in all directions uniformly. However, the distance between earth and sun is very large, and we may approximate the electromagnetic waves arriving at the surface of the earth as plane waves. For this type of wave, $\vec{E}$ and $\vec{B}$ are perpendicular. Thus

$$|\vec{S}| = \left| \frac{1}{\mu_{90}} \vec{E} \times \vec{B} \right| = EH = 1.35 \times 10^3\ W \cdot m^{-2}$$

where we have used the fact that $|\vec{H}| = \left|\frac{\vec{B}}{\mu_0}\right|$ in vacuum. (H is the magnetic field intensity.)

But in the electromagnetic field in vacuum,

$\varepsilon E^2 = \mu_0 H^2$, or $E\sqrt{\varepsilon_0/\mu_0} = \text{H}$. Then

$$\text{E} \times \sqrt{\varepsilon_0/\mu_0}\, E = EH = 1.35 \times 10^3\ \text{W} \cdot \text{m}^{-2}$$

or $$\text{E}^2 = \sqrt{\mu_0/\varepsilon_0} \times 1.35 \times 10^3\ \text{W}\,.\,\text{m}^{-2}$$

$$\text{E} = \sqrt{5.09 \times 10^5}\ \text{V}\,.\,\text{m}^{-1} = 0.71 \times 10^3\ \text{V}\,.\,\text{m}^{-1}.$$

Similarly,

$$\text{B} = \mu_0\text{H} = \frac{\mu_0\,(1.35 \times 10^3\ W\,.\,m^{-2})}{E}$$

$$\text{B} = \frac{(4\pi \times 10^{-}\ Weber\,.\,A^{-1}\,.\,m^{-1})\,(1.35 \times 10^3\ W\,.\,m^{-2})}{71 \times 10^3\ V\,.\,m^{-1}}$$

$$\text{B} = 2.39 \times 10^{-6}\,\frac{Weber\,.\,W\,.\,A^{-1}\,.\,m^{-2}}{V}$$

But 1 W = 1 J . s^{-1} and 1 V = 1 J . C^{-1} whence

$$\text{B} = 2.39 \times 10^{-6}\,\frac{Weber\ .\ J\ .\ s^{-1}\ .\,A^{-1}\ .\ m^{-2}}{J\ .\ C^{-1}}$$

$$\text{B} = 2.39 \times 10^{-6}\ \text{Weber} \cdot \text{m}^{-2}.$$

(b) The electromagnetic energy density (or, energy per unit volume) in an electromagnetic field in vacuum is $\mu_0 H^2 = \varepsilon_0 E^2$. The energy falling on 1 m^2 of the earth's atmosphere in 1 s is the energy initially contained in a cylinder 1 m^2 in cross section and 3×10^8 m in length; for all this energy travels to the end of the cylinder in the space of 1 s. Hence the energy density near the earth is

$$\mu_0 H^2 = \varepsilon_0 E^2 = \frac{1.35 \times 10^3\ W\,.\,m^{-2}}{3 \times 10^8\,m\,.\,s^{-1}}.$$

Here, ε_0 is the permittivity of free space.

$$E^2 = \frac{1.35 \times 10^3 \ W.m^{-2}}{8.85 \times 10^{-12} \ C^2.N^{-1}.m^{-2} \times 3 \times 10^8 \ m.s^{-1}}$$

$$E^2 = \frac{1.35 \times 10^7}{26.55} \ \frac{W}{C^2.N^{-1}.m.s^{-1}}$$

But $\quad 1 \ W = 1 \ J.s^{-1} = 1 \ N.m.s^{-1}$

$$E^2 = 5.085 \times 1^{-5} \ N/C \ N^{-1} = 5.085 \times 10^5 \ N^2/C^2$$

or $\quad E = .71 \times 10^3 \ N \cdot C^{-1} = .71 \times 10^3 \ V.m^{-1}$

Also $\quad \mu_0 H^2 = \dfrac{B^2}{\mu_0} = \dfrac{1.35 \times 10^3 \ W.m^{-2}}{3 \times 10^8 \ m.s^{-1}}$

or $\quad B^2 = \dfrac{4\pi \times 10^{-7} \ N.A^{-2} \times 1.35 \times 10^3 \ W.m^{-2}}{3 \times 10^8 \ m.s^{-1}}$

$$B = 2.36 \times 10^{-6} \ Wb.m^{-2}.$$

Problem 13. Show that a particle subjected to two simple harmonic vibrations of the same frequency, at right angles and out of phase, traces an elliptical path which degenerates to two coincident straight lines if the phase difference is π. Indicate the relevance of this to a half-wave plate.

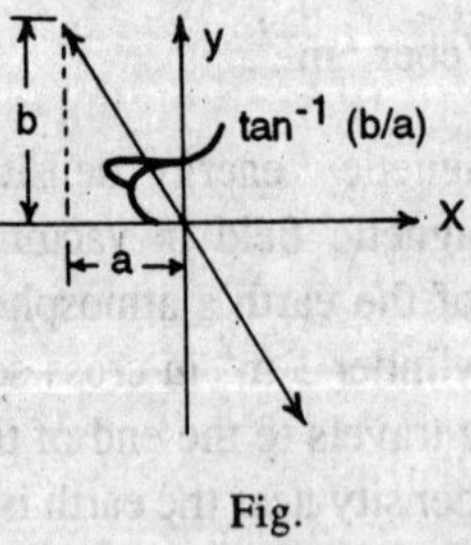

Fig.

Solution: Let the vibrations be taking place along the x- and y-axes with a phase difference of ϕ between them. Then if $x = a \sin \omega t$, $y = b \sin(\omega t + \phi)$. Since $\sin(\alpha + \beta) = \sin \alpha \cos \beta + \cos \alpha \sin \beta$.

$$\frac{y}{b} = \sin \omega t \cos \phi + \cos \omega t \sin \phi$$

$$= \frac{x}{a} \cos \phi + \sqrt{1 - \frac{x^2}{a^2}} \cdot \sin \phi$$

Here we have used the fact that

$$\cos \omega t = \sqrt{1 - \sin^2 \omega t}$$

$$\frac{y^2}{b^2} + \frac{x^2}{a^2} \cos^2 \phi - \frac{2xy}{ab} \cos \phi = \left(1 - \frac{x^2}{a^2}\right) \sin^2 \phi$$

$$\frac{x^2}{a^2} (\cos^2 \phi + \sin^2 \phi) + \frac{y^2}{b^2} - \frac{2xy}{ab} \cos \phi = \sin^2 \phi$$

or $$\frac{x^2}{a^2} + \frac{y^2}{b^2} - \frac{2xy}{ab} \cos \phi = \sin^2 \phi$$

This is the general equation of an ellipse where the major and minor axes do not coincide with the x- and y-axes. Thus the particle always has x- and y-coordinates such that the point they define lies on an ellipse. The particle thus follows an elliptical path.

If $f = \pi/2, 3\pi/2, 5\pi/2, \ldots$, the equation of the path reduces to $(x^2/a^2) + (y^2/b^2) = 1$, which is an ellipse with the major and minor axes coincident with the coordinate axes.

When $\phi = \pi$, the equation of the path becomes

$$\frac{x^2}{a^2} + \frac{y^2}{b^2} + \frac{2xy}{ab} = 0,$$

that is $$\left(\frac{x}{a} + \frac{y}{b}\right)^2 = 0.$$

This is the equation of two coincident straight lines $x/a = -y/b$, inclined to the negative x-axis at an angle $\tan^{-1} (b/a)$. (See figure.)

In the case of a half-wave plate, plane-polarized light striking the plate is split up in two components, O and E, plane-polarized at right angles to one another and initially in phase. These pass through the plate at different speeds and the thickness is such that on emergence the two beams are out of phase by π. Any particle affected by the two components will thus be affected by two simple harmonic vibrations at right angles, out of phase by π. As can be seen from the above analysis, the particle would trace a straightline path. This means that the two components are equivalent to a single vibration at an angle $\tan^{-1}(b/a)$ to the slower component, (b/a) being the ratio of the amplitudes of the components of the incident light on entering the plate. If the plane-polarized light is striking the plate at an angle of 45° to the two transmission directions, then it is resolved into two equal components so that $b = a$. The emerging light is thus plane-polarized in a direction making an angle of – 45° with each of the principal directions in the plate.

Problem 14. The captain of a submarine fitted with a directional transmitter finds that he receives bad echoes from the sea bed when he is submerged if the transmission direction makes an angle greater than 45° with the vertical (sea Figure A). He wishes to transmit a message to shore using radiation which is completely horizontally polarized. How far from the coast should he surface if the receiving point is a house on top of a coastal cliff 500 ft high and he intends to polarize his radiation by reflection on the sea surface? Take the location of the transmitter as 12 ft above the water when the submarine has surfaced.

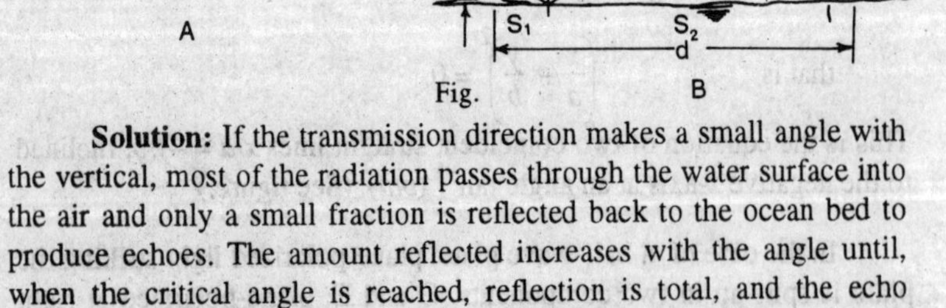

Fig.

Solution: If the transmission direction makes a small angle with the vertical, most of the radiation passes through the water surface into the air and only a small fraction is reflected back to the ocean bed to produce echoes. The amount reflected increases with the angle until, when the critical angle is reached, reflection is total, and the echo becomes troublesome. The refractive index of water for the radiation used is found from Snell's Law,

$$n \sin \theta_i = n_{air} \sin \theta_r$$

where θ_i and θ_r are the angles of incidence and refraction for the transmitted signal. We are told that the critical angle is $\theta_i = 45^\circ$. At the critical angle $\theta_r = 90^\circ$. Hence,

$$n = \frac{1}{\sin\ 45^\circ} = \sqrt{2} = 1.414.$$

When the submarine has surfaced, it can produce a completely plane-polarized beam by reflection of the radiation at the Brewster angle from the surface of the sea. The angle required is tan $f = n =$ 1.414. That is, $f = 54.75^\circ$. (See figure (B)).

Thus, the distance of the submarine from the cliff is

$$d = s_1 + s_2$$

From figure (B), however,

$$\frac{12\,ft}{s_1} = \tan\,(90 - \phi). = \tan\,35.25^\circ$$

and $$\frac{500\,ft}{s_2} = \tan\,(90 - \phi) = \tan\,35.25^\circ$$

or $$d = \frac{500ft}{\tan\,35.25^\circ} + \frac{12ft}{\tan\,35.25^\circ} = \frac{512}{0.707}\text{ ft} = 724\text{ ft}.$$

VIBRATING RODS & STRINGS

Problem 15. A string 100 centimetre long has a frequency of 200 vibrations per second. What is the frequency of a similar string under the same tension but 50 centimetres long?

Solution: Since the frequency is inversely proportional to length, the frequency of the 50-centimetres string is 2×200, or 400 vps.

Problem 15a. A string 100 centimetres long has a mass of 4 grams and emits a frequency of 250 vibrations per second. Another string 100 centimetres long has a mass of 1 gram. What is the frequency of the second string if it is put under the same tension as the first?

Solution. The second string has one-fourth the mass of the first. The square root of ¼ is ½ . Since the frequency varies inversely as the square root of the mass per unit length, the frequency of the second string is 2 × 250 or 500 vps.

Problem 16. A brass rod of density 8.5 g . cm^{-3} and length 100 cm is clamped at the centre. When set into longitudinal vibration it emits a note two octaves above the fundamental note emitted by a wire also of 100cm length weighing 0.295 g and under a tension of 20 kg weight which is vibrating transversely. What is Young's modulus for brass?

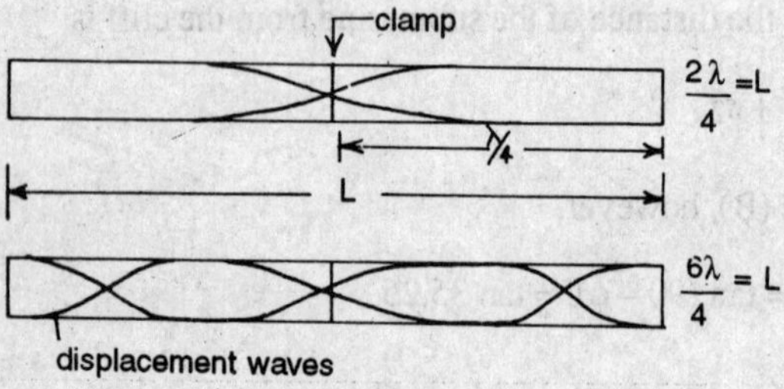

Fig.

Solution: The rod vibrates with its centre clamped (see figure). Its fundamental frequency of vibrations is thus such that the centre of the rod is a node and each of the ends an antinode. The length of the rod, L, is half a wavelength (λ). Thus l = 2L. Further, the speed of sound in the rod is $c = \sqrt{Y/\rho}$, where Y is Young's modulus for brass and ρ its density. Hence the frequency of vibration is

$$f = \frac{c}{\lambda} = \frac{1}{2L}\sqrt{\frac{Y}{\rho}}$$

For the vibrating wire and length is the same and if μ is the mass per unit length of the wire ($m = m/L$) the frequency of the fundamental vibration is

$$f_1 = \frac{1}{2L}\sqrt{\frac{S}{\mu}}\ \frac{1}{2L}\sqrt{\frac{SL}{m}},$$

where m is the mass of the wire. But $f = 4f_1$, since on frequency is two octaves above the other. Hence

$$\frac{1}{2L}\sqrt{\frac{Y}{\rho}} = \frac{1}{2L}\sqrt{\frac{SL}{m}}$$

or $$Y = \frac{16\rho SL}{m}$$

$$= \frac{16 \times 8.5\ g\,.\,cm^{-2} \times 2 \times 10^4\ kg \times 981\ cm\,.\,s^{-2} \times 100\ cm}{0.295 g}$$

$= 9.04 \times 10^{11}$ dynes $\cdot$ cm^{-2}.

Problem 17. A string under a tension of 256 newtons has a frequency of 100 vibrations per second. What is its frequency when the tension is changed to 144 newtons?

Solution: The tension is only $\frac{144}{256}$ as much as it was at first. The square root of $\frac{144}{256}$ is $\frac{12}{16}$, or $\frac{3}{4}$. Since the frequency is directly proportional to the square root of the tension, the new frequency is

$$\frac{3}{4} \times 100, \text{ or } 75 \text{ vps.}$$

Problem 18. One end of a horizontal wire is fixed and the other passes over a smooth pulley and has a heavy body attached to it. The frequency of the fundamental note emitted when the wire is plucked is 392 cycles $\cdot$ s^{-1}. When the body is totally immersed in water, the frequency drops to 343 cycles $\cdot$ s^{-1}. Calculate the density of the body.

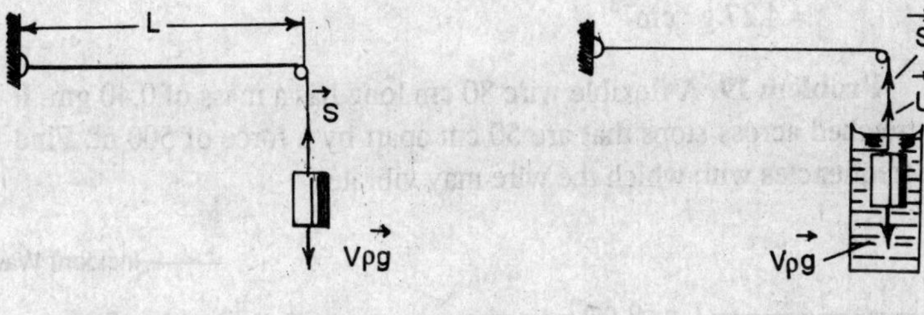

Fig.

Solution: Let the density of the body be ρ and its volume be V. The density of water is 1 g . cm^{-3} = ρ_0. In the first case, the weight of the body is balanced by the tension S in the wire. In the second case, a third force, the buoyancy, enters into the calculation. (See the figure).

The weight of the body is balanced partly by the new tension in the wire, $\vec{S_0}$, and partly by the buoyancy, $\vec{U}$, acting on it according to Archimedes' principle. That is, $\vec{U}$ is equal to the weight of water displaced by the body. Since the volume of the displaced water is equal to the volume of the body, then U = ($V\rho_0$) g where $V\rho_0$ is the mass of the displaced water. Thus

$$S = V\rho g$$

where $V\rho$ is the mass of the body and

$$S_0 + U = S_0 + V\rho_0 g = V\rho g,$$

the buoyancy due to the air in the first case being ignored.

The frequencies of the fundamental notes emitted in the two cases are

$$f_1 = \frac{1}{2L}\sqrt{\frac{S}{\mu}} \quad \text{and} \quad f_0 = \frac{1}{2L}\sqrt{\frac{S_0}{\mu}}$$

where μ is the mass of the wire per unit length (the mass density). Thus

$$\frac{f_0^2}{f_1^2} = \frac{S_0}{S} = \frac{V\rho g - V\rho_0 g}{V\rho g} = \frac{\rho - \rho_0}{\rho}$$

$$r = \frac{f_1^2\,\rho_0}{f_1^2 - f_0^2} = \frac{392^2\ s^{-2} \times 1\ g\,.\,m^{-3}}{(3{,}92^2 - 343^2)^{-2}}$$

$$= 4.27\ \text{g} \cdot \text{cm}^{-3}$$

Problem 19. A flexible wire 80 cm long has a mass of 0.40 gm. It is stretched across stops that are 50 cm apart by a force of 500 nt. Find the frequencies with which the wire may vibrate.

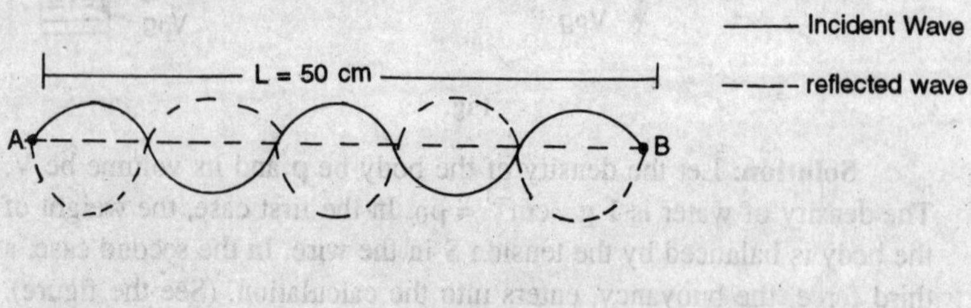

Fig.

Solution: The speed of a wave through a medium is determined by its elasticity and inertia. The elasticity is what causes a restoring force to act on any part of the medium displaced from its equilibrium position. The reaction of the displaced portion of the medium to the restoring force depends on its inertia. For a stretched wire, the tension T is a measure of its elasticity; the greater the tension, the greater is the elastic restoring force on the displaced portion of wire. The inertia is measured by *m*, the mass per unit length of the wire. The speed of the wave has been found both analytically and experimentally to be,

$$v = \sqrt{\frac{T}{\mu}} = \sqrt{\frac{500\,nt}{0.40 \times 10^{-3}\ kg/080\,m}}$$

$$= \sqrt{10^6\ m^2/sec^2} = 1000\ \text{m/sec.}$$

The wire can only vibrate in a standing wave pattern, since it is stopped at two points (see figure). The incident wave is reflected at points A and B and reinforces the later incident waves, producing the pattern shown. Hence, λ is related to L by,

$$\mathrm{L} = \frac{n\lambda}{2} \qquad (1)$$

Hence,

$$\lambda = \frac{2L}{n}$$

Since the velocity of a sound wave is related to its frequency f and wavelength by

$$v = f$$

we obtain,

$$\frac{v}{f} = \frac{2L}{n}$$

or,

$$f = \frac{nv}{2L} = \frac{n\,(10^3\ m/s)}{(2)\ (50 \times 10^{-2}\ m)}$$

$$f = n \times 10^3\ s^{-1}$$

Putting in values of *n*, we find,

$$f_1 = 10^3\ s^{-1}$$

$$f_2 = 2 \times 10^3\ s^{-1}$$

$$f_3 = 3 \times 10^3\ s^{-1}$$

. .

. .

$$f_n = n \times 10^3\ s^{-1}$$

The other possible frequencies are the integral multiples of 1000 vibrations/sec, that is, 2000, 3000, 4000, ... , vibrations/sec.

Problem 20. To what tension must a brass wire of cross-sectional area 10^{-2} cm^2 be subjected so that the speed of longitudinal and transverse waves along it may be the same? Young's modulus for brass is 9.1×10^{11} dynes · cm^{-2} Is this situation physically realizable?

Solution: The speed of transverse waves in the wire is $\sqrt{S/\mu}$, and of longitudinal waves $\sqrt{Y/\rho}$, where Y is the Young's modulus of brass, S is the tension in the wire, ρ is the density of brass, and *m* is the mass per unit length of the wire.

In this problem we require that $\sqrt{S/\mu} = \sqrt{Y/\rho}$.

$$\therefore \qquad S = \frac{\mu}{\rho} Y = \frac{\text{mass per unit length}}{\text{mass per unit volume}} Y = AY$$

where A is the cross-sectional area of the wire. Therefore

$$S = 10^{-2}\ \text{cm}^2 \times 9.1 \times 10^{11}\ \text{dynes} \cdot \text{cm}^{-2}$$

$$= 9.1 \times 10^9\ \text{dynes}$$

Since the formula for Young's modulus is Y = (S/A) ($\Delta l/l$), then S = AY implies $\Delta l/l = 1$. In other words, the wire must stretch by an amount equal to its length *l*. The elastic limit would have been passed long before this point. The situation is therefore physically unrealization and longitudinal waves will always travel faster than transverse ones in the wire.

Problem 21. Two identical wires are stretched by the same tension of 100 N, and each emits a note of frequency 200 cycles . s^{-1}. The tension in one wire is increased by 1 N. Calculate the number of beats heard per second when the wires are plucked.

Solution: The frequency of the fundamental note emitted by each wire before the tension change occurs is

$$\upsilon = \frac{1}{2L}\sqrt{\frac{T}{\mu}} \qquad (1)$$

If T changes, v will also change. We can find the relation between these 2 changes by taking the derivative of (1) with respect to T

$$\frac{d\upsilon}{dT} = \frac{1}{2L}\left(\frac{1}{2}\left(\frac{T}{\mu}\right)^{-1/2}\frac{1}{\mu}\right)$$

$$\frac{d\upsilon}{dT} = \frac{1}{4L}\sqrt{\frac{\mu}{T}}\,\frac{1}{\mu}$$

$$\frac{d\upsilon}{dT} = \frac{1}{4LT}\sqrt{\frac{T^2\mu}{T\mu^2}} = \frac{1}{4LT}\sqrt{\frac{T}{\mu}}$$

From (1)

$$\frac{d\upsilon}{dT} = \frac{\upsilon}{2T}$$

Hence, $$\Delta\upsilon \approx \frac{\upsilon}{2}\frac{\Delta T}{T}$$

where $\Delta\upsilon$ is the frequency difference induced in the string as a result of a change in tension ΔT. In other words, Δv is the number of beats observed if the string's tension is changed by an amount ΔT. Using the given data

$$\Delta\upsilon = \left(\frac{200}{2}\ cycles\ .\ s^{-1}\right)\left(\frac{1N}{100N}\right)$$

$$\Delta\upsilon = 1\ \text{cycle}\cdot\text{s}^{-1}$$

Problem 22. Calculate the velocity of a transverse pulse in a string under a tension of 20 1b if the string weighs 0.003 1b/ft.

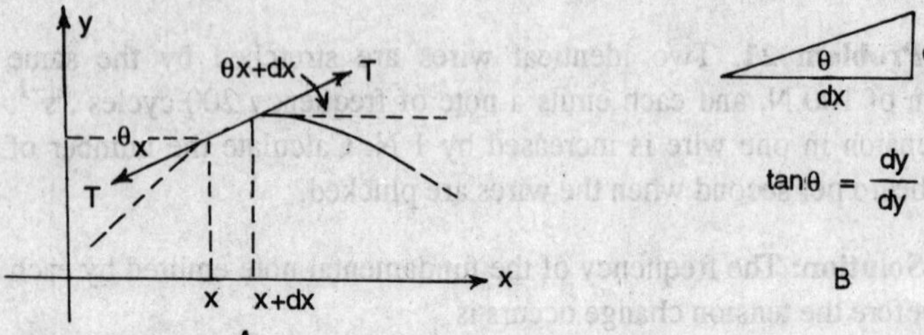

Fig.

Solution: Consider the section of string shown in figure A, which is under the tension T. This segment can be interpreted as being part of the pulse being transmitted along the string. The net force acting in the *y* direction is

$$F_y = T \sin \theta_{x+dx} - T \sin \theta x$$

Since θ_x and θ_{x+dx} are small, we have approximately,

$$\sin \theta x \approx \theta x \approx \tan \theta x$$

$$\sin \theta_{x+dx} \approx \theta_{x+\delta x} \approx \tan \theta_{x+dx}$$

Therefore, $F_y = T(\tan \theta_{x+dx} - \tan \theta_x)$

but $\tan \theta_{x+dx} = \left(\frac{\partial x}{\partial x}\right) x + dx$

$$\tan \theta_x = \left(\frac{\partial x}{\partial x}\right) x \qquad \text{(See fig. B)}$$

∂y and ∂x have the same meaning as *dy* and *dx*, they are small increments of *y* and of *x*, respectively. The ∂ symbol indicates that more than one variable is under consideration.

$$\text{Hence, } F_y = T\left(\left(\frac{\partial y}{\partial x}\right)_{x+dx} - \left(\frac{\partial y}{\partial x}\right)_x\right) \qquad (1)$$

Now, by Taylor's theorem, given a continuous and differentiable function G(x),

$$G(x + dx) = G(x) + \frac{\partial G}{\partial x}\, dx + \ldots$$

$$G(x + dx) - G(x) \approx \frac{\partial G}{\partial x}\, dx \qquad (2)$$

Higher order terms have been neglected because dx is so small that terms involving $(dx)^2$ or higher must necessarily be negligible. If $G(x) = \partial y/\partial x$, then we have from (2)

$$\left(\frac{\partial y}{\partial x}\right)_{x+dx} - \left(\frac{\partial y}{\partial x}\right)_{x} = \frac{\partial}{\partial x}\left(\frac{\partial y}{\partial x}\right) dx = \frac{\partial^2 y}{\partial x^2}\, dx$$

Hence, (1) becomes $F_y = T\dfrac{\partial^2 y}{\partial x^2}\, dx.$

Denote the mass per unit length of the string by μ (i.e. $\mu = dm/dx$). The mass of the segment dx is then $m = \mu dx$. By Newton's second law, the vertical acceleration of the string segment, a_y, is

$$F_y = ma_y$$

$$T\frac{\partial^2 Y}{a x^2}\, dx = (\mu dx)\frac{\partial^2 y}{\partial t^2}$$

$$\frac{\partial^2 y}{\partial x^2} = \frac{\mu}{T}\frac{\partial^2 y}{\partial t^2} \qquad (3)$$

Equation (3) is of the form of the wave equation

$$\frac{\partial^2 y}{\partial x^2} = \frac{1}{v^2}\, \frac{\partial^2 y}{v^2\, \partial t^2}$$

where $y(x, t)$ is the displacement of the wave at position x and at time t, and v is the velocity of the wave. For the transverse wave, then

$$v = \sqrt{\frac{T}{\mu}}$$

Returning to our original problem,

T = 20 1b,

$$\mu = \frac{0.003}{32} \frac{1b/ft}{ft/sec^2},$$

$$v = \sqrt{\frac{20\ lb \times 32 ft/sec^2}{0.003\ 1b/ft}} = 461 \frac{ft}{sec}.$$

TRAVELLING AND STANDING WAVES

Problem 23. (a) The maximum pressure variation P that the ear can tolerate in loud sounds is about 28 N/m^2 (=28 Pa). Normal atmospheric pressure is about 100,000 Pa. Find the corresponding maximum displacement for a sound wave in air having a frequency of 1000 Hz.

(b) In the faintest sound that can be heard at 1000 Hz the pressure amplitude is about 2.0×10^{-5} Pa. Find the corresponding displacement amplitude.

Solution: (a) The pressure amplitude is given by

$$y_m = \frac{P}{\kappa \rho_0 v^2}$$

The speed of sound in air is, $v = 331$ m/s so that

$$k = \frac{2\pi}{\lambda} = \frac{2\pi\upsilon}{v} = \frac{2\pi \times 10^3}{331} m^{-1} = 19 \text{ m}^{-1}.$$

The density of air ρ_0 is 1.22 kg/m^3. Hence, for P = 28 Pa

$$y_m = \frac{28}{(19)(1.22)(331)^2} \text{ m} = 1.1 \times 10^{-5} \text{m}.$$

The displacement amplitude for the loudest sounds are about 10^{-5}m, a very small values.

(b) From $y_m = P/k\rho s v^2$, using these values for k, v, and ρ_0, with P = 2.0×10^{-5} N/m^2,

$$y_m = 8 \times 10^{-12} m = 10^{-11} \text{ m}.$$

Problem 24. A transverse sinusoidal wave is generated at one end of a long horizontal string by a bar whch moves the end up and down through a distance of 0.50 cm. The motion is continuous and is repeated regularly 120 times per second.

(a) If the string has a linear density of 0.25 kg/m and is kept under tension of 90 N, find the speed, amplitude, frequency, and wavelength of the wave motion.

(b) Assuming the wave moves in the +x-direction and that, at t = 0, the end of the string described by x=0 is in its equilibrium position y = 0, write the equation of the wave.

(c) As this wave passes along the string, each particle of the string moves up and down at right angles to the direction of the wave motion. Find the velocity and acceleration of a particle 2.0 ft. from the end.

Solution: (a) The end moves 0.25 cm away from the equilibrium position, first above it, then below it; therefore, the amplitude ym is 0.25 cm.

The entire motion is repeated 120 times each second so that the frequency is 120 vibrations per second, or 120 Hz.

The wave speed is given by $v = \sqrt{F/\mu}$. But F = 90 N and μ = 0.25kg/m, so that

$$v = \sqrt{\frac{90N}{0.25kg/m}} = 19 \text{ m/s.}$$

The wavelength is given by $\lambda = v/\upsilon$, so that

$$\lambda = \frac{19\ m/s}{120\ vib/s} = 16 \text{ cm.}$$

(b) The general expression for a transverse sinusoidal wave moving in the +x-direction is

$$y = y_m \sin(kx - \omega t - \phi).$$

Requiring that $y = 0$ for the conditions $x = 0$ and $t = 0$ yields

$$0 = y_m \sin(-\phi),$$

which means that the phase constant ϕ may be taken to be zero. Hence for this wave

$$y = y_m \sin(kx - \omega t).$$

and with the values just found,

$$y_m = 0.25 \text{ cm},$$

$$\lambda = 16 \text{ cm} \qquad \text{or} \qquad k = \frac{2\pi}{\lambda} = \frac{2\pi}{16cm} = 0.39 \text{ cm}^{-1},$$

$$v = 19\text{m/s} = 1900 \text{ cm/s or } \omega = vk = (1900 \text{ cm/s})(0.39 \text{ cm}^{-1})$$

$$= 740 \text{ s}^{-1} = 740 \text{ Hz},$$

the equation for the wave is

$$y = 0.25 \sin(0.39x - 740t)$$

where x and y are in centimetres and t is in seconds.

(c) The general form of this wave is

$$y = y_m \sin(kx - \omega t) = y_m \sin k(x - vt).$$

The v in this equation is the constant horizontal velocity of the wavetrain. It is not the velocity of a particle in the string through which this wave moves; this particle velocity is neither horizontal nor constant. Each particle moves vertically, that is, in the y-direction. In order to determine the particle velocity designated by the symbol u, concentrate on a particle at a particular position x-that is, x is now a constant in this equation-and see how the particle displacement y changes with time. With x constant

$$u = \frac{\partial y}{\partial t} = -y_m\omega \cos(kx - \omega t),$$

in which the partial derivative $\partial y/\partial t$ shows that although in general y is a function of both x and t, assume that x remains constant so that t becomes the only variable. The acceleration a of the particle at this (constant) value of x is

$$a = \frac{\partial^2 y}{\partial t^2} = \frac{\partial u}{\partial t} = -y_m\omega^2 \sin(kx - \omega t) = -\omega^2 y.$$

For a particle at $x = 62$ cm with

$$y_m = 0.25 \text{ cm}, \qquad k = 0.39 \text{ cm}^{-1}, \qquad w = 740 \text{ s}^{-1},$$

$$u = -\, y_m v \cos (kx - \omega t)$$

or $u = -\,0.25\,(740) \cos \left((0.39)(62) - 740\, t\right) = -\,185 \cos (24 - 740t)$

and $$a = -\omega^2 y$$

or $a = -\,(740)^2\, 0.25 \sin [(0.39)\,(62) - (740)\, t] = -\,13.7 \times 10^4 \sin (24 - 740\, t)$

where t is expressed in seconds u in cm/s and a in cm/s^2.

3

Sound

Problem 1. What is the frequency of a 2-cm sound wave in sea water? The velocity of sound in sea water is $v = 1.53 \times 10^5$

Solution: The 2-cm is the wavelength λ of the sound wave. The frequency is given by

$$\nu = \frac{\upsilon}{\lambda}$$

$$= \frac{1.53 \times 10^5 \; cm/sec}{2cm}$$

$$= 7.6 \times 10^4 \text{ Hz} = 76 \text{ kHz}$$

which is an ultrasonic wave (that is, above the human audible range). A 2 cm sound wave in air would be audible.

Problem 2. The frequency of middle C is 256 sec^{-1}. What is its wavelength in air?

Solution: The velocity of sound in air varies with temperature. At 18°C, which is a common room temperature it is 3.4×10^4 cm/sec. Hence, since

$$\text{Wave velocity} = \text{Frequency} \times \text{Wavelength}$$

we see that

$$\text{Wavelength} = \frac{\text{Wave velocity}}{\text{Frequency}}$$

$$= \frac{3.4 \times 10^4 \; cm/sec}{256 \; sec^{-1}}$$

$$= 133 \text{ cm.}$$

This is approximately 4 feet. When the pianist plays middle C, the high pressure regions travelling toward a listener through the air are about 4 feet apart, but they are moving with the very high speed of 3.4×10^4 cm/sec (760 mph) and so 256 of them arrive at the listener's ear every second.

Problem 3. A sound with a frequency of 1,000 cycles per second is produced when the air temperature is 15° centigrade. What is the wavelength of this sound?

Solution: Velocity (v), frequency (n) and wavelength (λ) are related by the relation $v = n\lambda$. The speed of sound at sea level at 0°C is 1090 ft/sec. This speed increases by 2 ft/sec for every degree centigrade above 0°C. Hence,

$$\text{Velocity in ft/sec} = 1090 + (2 \times 15)$$

$$= 1090 + 30$$

$$= 1120$$

Since $v = n\lambda$,

$$\lambda = v/n$$

$$= \frac{1120\, ft/sec}{1000\, cycles/sec}$$

$$= 1.12 \text{ ft/cycle}$$

Problem 4. Find the theoretical speed of sound in hydrogen at 0°C. For a diatomic gas $r = 1.40$, and for hydrogen M = 2.016 gm/mole.

Solution: Sound waves are longitudinal mechanical waves. That is, they are propagated in matter and the particles transmitting the wave oscillate in the direction of propagation of the wave. The speed of the wave is determined by the elastic and inertial properties of the medium. For a gaseous medium, the elastic property depends on the undisturbed gas pressure p_0 and the inertial property on l_0, the density/mole of the gas. With r a constant called the ratio of specific heats for the gas, the velocity is

$$v = \sqrt{r p_0/\rho_0}$$

Since for one mole of gas,

$$p_0 = \frac{RT}{V} = \frac{RT\rho_0}{M}$$

we have $v = \sqrt{\frac{RT}{M}}$

$$= \sqrt{\frac{1.40\ [8.317\ joules/(mole K^{\circ})]\ (273^{\circ} K)}{2.016 \times 10^{-3}\ kg/mole}}$$

$$= \sqrt{\frac{1.40 \times 8.317 \times 273\ joules}{2.016 \times 10^{-3}\ kg}}$$

$$= 1.25 \times 10^3 \text{ m/sec.}$$

Problem 5. A sound is returned as an echo from a distant cliff in 4 sec. How far away is the cliff, assuming the velocity of sound in air to be 1100 ft/sec?

Solution: The sound travels the distance (s) between cliff and observer twice. Hence

$$2s = v \times t$$

where v is the velocity of sound, and t is the time required for the entire trip. Therefore,

$$s = \frac{v \times t}{2} = \frac{(1100\ ft/sec) \times (4\ sec)}{2} = 2200 \text{ ft.}$$

Problem 6. Using the speed of sound at sea level as approximately 750 miles per hour, what is the velocity of a jet plane travelling at Mach Number 2.2?

Solution:

$$\text{Mach Number} = \frac{\text{velocity of body}}{\text{velocity of sound}}, \qquad \text{and}$$

velocity of body = Mach Number × velocity of sound, whence

velocity of sound = 750 mph × 2.2

= 1650 mph = 2200 ft/sec approximately.

Problem 7. Compute the speed of sound in the steel rails of a railroad track. The weight density of steel is 490 lb/ft^3, and Young's modulus for steel is 29×10^6 1b/in^2.

Solution: For an elastic medium, the speed of longitudinal waves is given by

$$v = \sqrt{\frac{Y}{\rho}} = \sqrt{\frac{Yg}{D}}$$

where Y is Young's modulus, ρ is the density of the medium and

$\rho = \frac{D}{g}$ where D is the weight density of the medium.

$$v = \sqrt{\frac{(29 \times 10^6 \ 1b/in^2)\ (32\, ft/s^2)}{490\ 1b/ft^3}}$$

In order to keep all length dimensions consistent, we must change 29×10^6 1b/in^2 to 1b/ft^2, Hence,

$$29 \times 10^6 \text{ 1b/in}^2 = \frac{29 \times 10^6 \ lb}{\frac{1}{44} ft^2}$$

$$29 \times 10^6 \quad \text{1b/in}^2 = 29 \times 144 \times 10^6 \text{ 1b/ft}^2$$

Therefore,

$$v = \sqrt{\frac{(29 \times 10^6 \times 144\ lb/ft^2)\ (32\, ft/sec^2)}{490\ 1b\ lb/ft^3}}$$

$$= 1.6 \times 10^4 \text{ ft/sec}$$

Problem 8. A small sports arena is designed by an architect in the form of a dome with radius of curvature R = 115 ft mounted on a cylindrical base 75 ft in radius and 30 ft in height.

The dome acts as a spherical mirror with a focal length *f* = 1/2R =

57.5 ft. The top of the dome is the vertex of the mirror. It is of interest to calculate the location of the focal point with respect to the ground surface of the arena.

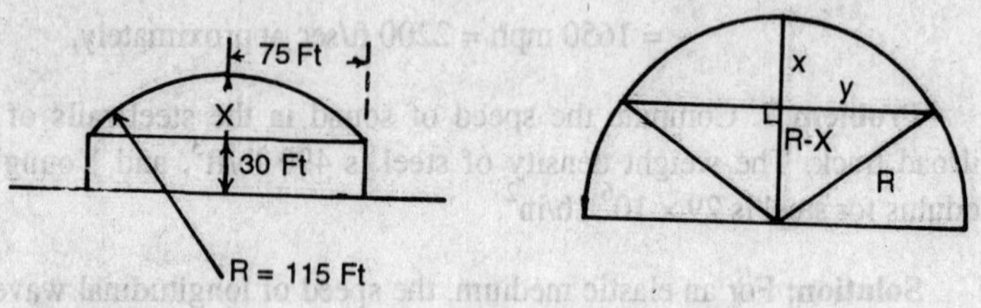

Fig.

Solution: From the diagram

$$y^2 + (R - x)^2 = R^2$$

$$(R - x)^2 = R^2 - y^2$$

$$= (115^2 - 75^2)\ ft^2$$

$$= 872\ ft^2$$

$$R - x = 87\ ft$$

$$x = (115 - 87)\ ft$$

$$x = 28\ ft$$

The distance from the vertex of the dome to the ground surface is $x + 30 = 58$ ft, the same as the focal length. Thus the focal point of the mirror lies on the ground surface at the centre of the arena.

As a result of this there will be a tendency for spectator noise to be focused at the centre of the ground surface, and the noise there is liable to be deafening. There exists a hockey arena that has been designed this way (accidentally), and in the centre ice region the noise is of such intensity that the players can not hear the whistles of the officials.

Problem 9. What is the wavelength of the sound wave emitted by a standard 440 cycles per second tuning fork?

Solution: Noting that the velocity (v), frequency (n), and wavelength (λ) of sound are related by $v = n\lambda$, and assuming the velocity of sound to be 34,000 cm/sec, or approx. 1100 ft/sec, we find

$$\lambda = v/n = \frac{1100ft}{440} = 2.5 \text{ ft (approx.)}$$

Problem 10. When two tuning forks are sounded simultaneously, a beat note of 5 cycles per second is heard. If one of the forks has a known frequency of 256 cycles per second, and if a small piece of adhesive tape fastened to this fork reduces the beat note to 3 cycles per second, what is the frequency of the other fork?

Solution: This problem involves the phenomenon of beats. When two similar waves are superimposed, the beat frequency represents the numerical difference in their frequencies. Hence, for the case in question,

$$n = (256 \pm 5) \text{ cycles/sec}$$

where n represents the unknown frequency.

It appears that n has two possible values, either 251 or 261. Now, when the standard fork is loaded with the tape, its frequency will decrease. Since the beat frequency is then reduced to 3 cycles per second, the unknown frequency must be less rather than more than 256. Hence,

$$n = 251.$$

Problem 11. Two trains moving along parallel tracks in opposite directions converge on a stationary observer, each sounding its whistle of frequency 350 cycles. s–1. One train is travelling at 50 mph. What must be the speed of the other if the observer hears 5 beats per second. The speed of sound in air is 750 mph.

Solution: When a source is moving toward a stationary observer, the latter hears a frequency for the emitted note which is related to the frequency of the source by the expression $f = uf_s\ (u - v_s)$ where u is the speed of sound in air, v_s is the speed of the source and f_s is the frequency of the sound emitted by the source. If one of the trains is moving toward the observer with v_1 and the other with speed v_2, then since both have the same frequency whistle,

$$f_1 = \frac{uf_s}{u - v_1} \quad \text{and}$$

$$f_2 = \frac{uf_s}{us - v_2} = \frac{750\ mph \times 30\ s^{-1}}{(750 - 50)\ mph} = 375 \text{ cycles} \cdot s^{-1}.$$

But the observer hears 5 beats per second. This corresponds to a frequency difference $f_1 - f_2 = \times 5$ cylcles . sec^{-1}. Hence, $f_1 = 370$ cycles $\cdot\ s^{-1}$ or 380 cycles . s^{-1}

$$\therefore f_1 = \frac{uf_s}{u - v_1} = 370\ s^{-1} \quad \text{or} \quad 380\ s^{-1}.$$

$$\therefore u - v_1 = \frac{750\ mph \times 350\ s^{-1}}{370\ s^{-1}} \quad \text{or}$$

$$\frac{750\ mph \times 350\ s^{-1}}{380\ s^{-1}} = 709.5 \text{ mph} \quad \text{or } 690.8 \text{ mph}$$

$$\therefore v_1 = 40.5 \text{ mph} \quad \text{or} \quad 59.2 \text{ mph.}$$

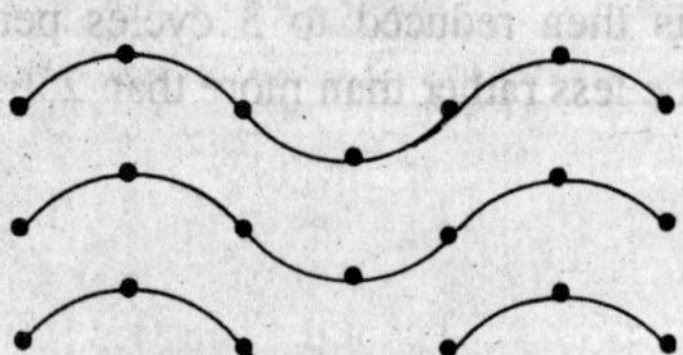

(a) The behavior of the atoms for a tranverse wave with a wavelength longer than the distance between atoms.

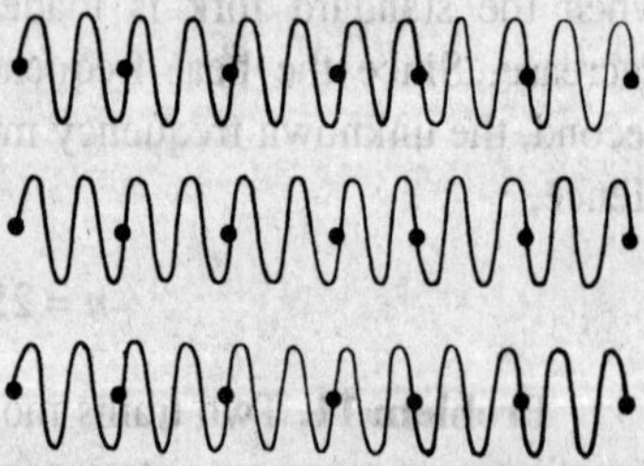

(b) There is no meaning in a wavelength shorter than the distance between atoms because there is nothing in between the atoms to oscillate.

Fig.

Problem 12. Estimate the upper limit to the frequency of "sound" waves in ordinary matter.

Solution: The relation $\nu = \frac{v}{\lambda}$, (1)

where *n*, *v*, and λ are the frequency, velocity and wavelength of the sound wave, respectively, makes it clear that the higher the frequency the shorter is the wavelength. We are therefore asking what is the shortest wavelength sound can have. The answer is that, when the wavelength becomes shorter than the distance apart of the atoms, the concept of a wave breaks down because there is nothing is between the atoms to oscillate (see figure). The atoms are closest together in solids and highly compressed gases.

However, when two atoms are about 10^{-8} cm apart they repel one another very strongly and it is difficult to force them any nearer. Consequently, in no material are the atoms or molecules found to be closer together than about 10^{-8} cm. This is therefore the order of magnitude of the shortest wavelength of sound.

The highest value found for the velocity of sound is about 10^6 cm/sec. Using equation (1).

$$\text{Highest frequency} = \frac{\text{Largest wave velocity}}{\text{Shortest wavelength}}$$

$$= \frac{10^6}{10^{-8}} \text{ approximately}$$

$$= 10^{14} \text{ sec}^{-1} \text{ approximately.}$$

This is much larger than the highest frequency yet produced experimentally, which is 2.5×10^{10} sec^{-1}.

The reader should note the character of these arguments, which are not precise calculations, but rather intelligent guesses. We have made it very plausible that the highest frequency is not very much bigger than 10^{14} sec^{-1}. It is conceivable that, by applying a very high pressure to the right gas or solid, we could force its atoms a little closer than 10^{-8} cm and achieve a velocity of sound a little larger than 106 cm/sec. In this way we might perhaps realize a frequency of, say 2×10^{14} sec^{-1}. We should be very surprised, though, if we ever pushed the frequency up to 10^{15} sec^{-1}. In fact we know that there is a limit to the procedure, because there is reason to believe that, when the pressure on a substance becomes extremely large, its atoms break up into separate electrons and nuclei.

Problem 13. What is the lowest frequency of the standing sound wave that can be set up between walls that are separated by 25 ft?

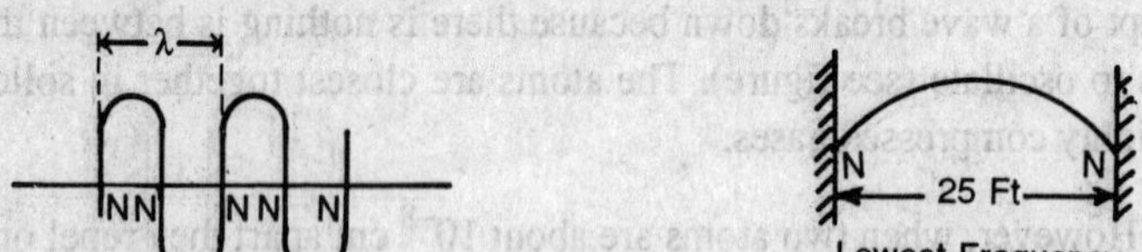

Illustration of Nodes

Lowest Frequency Standing Wage

Fig.

Solution: A. sinusoidal wave that maintains its overall shape between two termination points is called a standing wave. Its amplitude does change with time. For such a wave, the end points have zero amplitude at all times. Zero amplitude points are called nodes and occur every half wavelength.

The wave of lowest frequency has the longest wavelength. For a given distance L, the standing wave of longest wavelength has $\lambda = 2L$ with the only nodes occurring at the termination points. Therefore,

$$\lambda = 2L = 2 \times 25 \text{ ft} = 50 \text{ ft}$$

The frequency of the wave is

$$\nu = \frac{\upsilon}{\lambda} = \frac{1100\, ft/sec}{50\, ft}$$

$$= 22 \text{ Hz}$$

which is close to the lowest frequency that can be heard by a human ear. Therefore, a room somewhat larger than 25 ft is necessary in order to set up standing waves of the lowest audible frequency (for example, organ notes of 16 Hz).

Problem 14. A sonar device emits waves of frequency 40,000 cycles $\cdot$ s^{-1}. The velocities of the wave in air and water are 1100 ft . s^{-1}

and 4200 ft · st^{-1}, respectively. What are the frequency of the wave in air and the wavelengths in air and water?

Suppose that the device is fixed to the bottom of a ship. It emits a signal and the echo from the ocean bed returns 0.8 s later. What is the depth of the ocean at that point?

Solution: The frequency of the waves emitted is the same in air or water. The surrounding medium has no influence on the vibration mechanism. Since $\lambda = c/f$ where c is the wavespeed, and f and λ are the frequency and wavelength of the sound, respectively, we have (a) in air

$$\lambda = \frac{1100\, ft\, .\, s^{-1}}{4 \times 10^4\ s^{-1}} = 2.75 \times 10^{-2}\text{ ft}$$

and (b) in water

$$\lambda = \frac{4200\, ft\, .\, s^{-1}}{4 \times 10^4\, s^{-1}} = 10.50 \times 10^{-2}\text{ ft.}$$

Since the velocity of sound in water is 4200 ft . st^{-1} and the echo returns in 0.8 s after traversing $2d$, where d is the ocean depth at that point, we have $s = ct$ or $2d = 4200\text{ ft . s}^{-1} \times 0.8\text{ s.}$

$\therefore d = 1680$ ft.

Problem 15. Two small loudspeakers A and B vibrate in phase at 800 cps, and are set apart by 10 ft. The speakers emit spherical waves which obey the inverse square law with respect to intensity. The intensity variation (for distances away from the speakers greater than a few inches) is

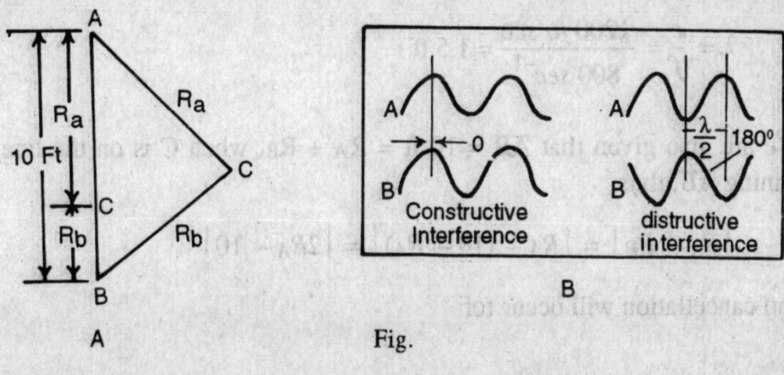

Fig.

$$I_A = \frac{60}{R_A^2} \qquad I_B = \frac{30}{R_B^2}$$

The units are arbitrary; R_A is the distance from speaker A, R_B is the distance from speaker B. Find the point of minimum intensity along the line joining the speakers. Find the first three points of zero intensity that are nearest to the speakers.

Solution: Assume that points of minimum and zero intensity occur only at points of destructive interference. Destructive interference of two waves occurs when the waves are 180° out of phase. The resultant amplitude at the point of interference will be the difference of the amplitude of the two waves at the point. The speed of sound is 1200 ft/sec. Consider point C (in the figure) such that

$$R_A = \overline{CA} \qquad R_B = \overline{CB}$$

Wave cancellation occurs when the two waves are 180° out of phase. Since the two waves originate with a 0° phase difference, one wave must undergo a 180° phase shift (relative to the second wave) for destructive interference to occur. In our case, the phase change is effected by letting the two waves move through different path lengths before they meet and interfere. Destructive interference will then occur when this path difference is an odd number of half wavelengths. This will insure a phase shift of 180°. Therefore,

$$|R_A - R_B| = (2n-1)\frac{\lambda}{2} \qquad n = 1, 2, 3, \ldots$$

Now, if λ and f are the wavelength and frequency of the sound wave, and c is its velocity.

$$\lambda f = c$$

$$\lambda = \frac{c}{f} = \frac{1200\, ft/sec}{800\, sec^{-1}} = 1.5 \text{ ft}$$

We are also given that $\overline{AB}$ = 10 ft = $R_A + R_B$, when C is on the line joining $\overline{AB}$; thus

$$|R_A - R_B| = |R_A - (10 - R_A)| = |2R_A - 10|$$

and cancellation will occur for

$$2RA - 10 = (2n - 1)\frac{\lambda}{2} \qquad n = 1, 2, 3, 4, \ldots$$

$$= (1)\frac{1.5}{2} = 0.75 \qquad \text{if } n = 1$$

$$= (3)\frac{1.5}{2} = 2.25 \qquad \text{if } n = 2$$

$$= (5)\frac{1.5}{2} = 3.75 \qquad \text{if } n = 3$$

$$= (7)\frac{1.5}{2} = 5.25 \qquad \text{if } n = 4 \text{ and so forth.}$$

The resultant intensity at the point of destructive interference is

$$I_C = |I_A - I_B|$$
$$= \left| \frac{60}{R_A^2} - \frac{30}{R_B^2} \right|$$

This follows from the very nature of destructive interference; the intensities of the interfering waves tend to try to cancel one another.

Problem 16. A vibrating tuning fork of frequency 384 cycles. s^{-1} is held over the end of a vertical glass tube, the other end of which dips into water. Resonance occurs when the top of the tube 21.9 cm and also 66.4 cm above the water surface. Calculate the speed of sound in air and end correction of the tube.

Assuming that the temperature is 13°C, what is the velocity of sound at 0°C (V_0)? (Fig. on next page)

Solution: Figure (C) shows the tuning fork in vibration. In order to send a compression wave down the tube, one tine of the fork is pulled back to position *c*, and then is let go. During one oscillation, the tine moves from *c* to *b* and then back to *c*. Now, the pipes with which we are dealing have water at one end. This end must then be a displacement node, for the air at this end of the column is always at rest. Alternatively, it is a pressure antinode. Similarly, the open end of the tube is a displacement maximum, or a pressure node.

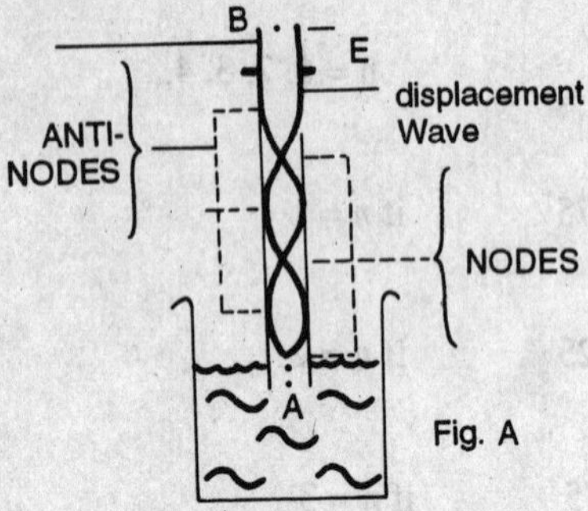

Fig. A

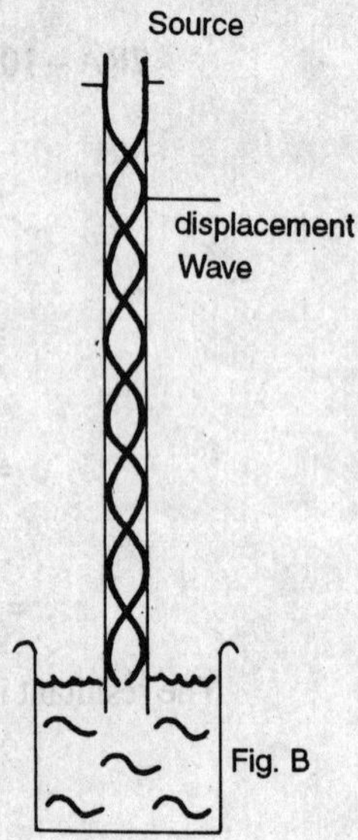

Fig. B

Fig. C

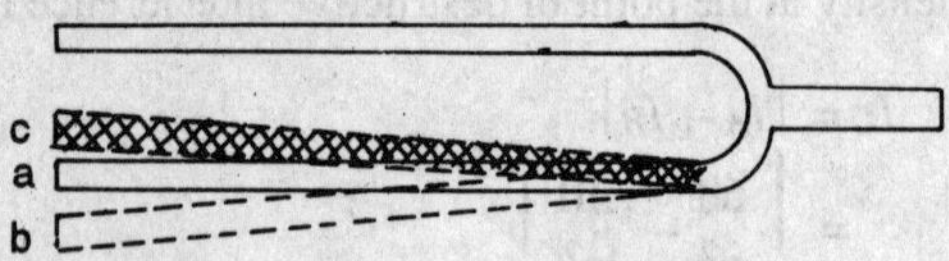

Fig.

Now, suppose the fork starts at point *c* and sends a compression wave down the tube. When this wave reaches the bottom of the tube, it must be reflected as a compression if that end is to remain a pressure antinode. The reflected compression travels back up the tube. If resonance is to occur, the compression must reach the tine of the tuning fork when the latter is ready to move from *b* to *c*. (In this situation, the tine is preparing to send a rarefaction down the tube). If this does not occur, the top of the tube will not remain a pressure node, for the reflected pressure wave and the new pressure wave will not cancel. Hence, in ½ oscillation of the tuning fork, the wave travels a distance 2L (twice the tube length). In general, resonance will occur if the wave travels a distance 2L in an odd number of half oscillations of the fork. If the wave velocity is v and the frequency of the fork is ν_{source}, the resonance condition is

$$2L = v(n + ½)/\nu_{source} \qquad (n = 0, 1, 2 \ldots) \qquad (1)$$

Since the wave frequency equals the source frequency, we may write

$$\upsilon_{wave} = \upsilon_{source} = \upsilon$$

But

$$\upsilon_{wave}\,\lambda = v$$

where λ is the wavelength of the wave. Hence, (1) becomes

$$2L = \frac{v\,(n + ½)}{\upsilon}$$

Solving this for v

$$v = \frac{2L\upsilon}{(n + 1/2)} \qquad (n = 0, 1, 2, \ldots) \qquad (2)$$

Therefore, we may measure v if we know the tube length, the source frequency, and the value of n. In practice, the tube is put in the water, and slowly drawn out as the fork vibrates. When the first resonance occurs, $n = 0$, and the length of tube extending out of the water is measured. Again, the tube is drawn out of the water. When the second resonance occurs, $n = 1$, and the tube length is again measured, etc. Each resonance will give us the same value for v, as equation (2) indicates.

Using the relation

$$v = \upsilon\lambda$$

where λ is the wavelength of the wave, we may also write (2)

$$v = \frac{2L}{(n + ½)}\,\frac{v}{\lambda}$$

or $$L = (n + ½)\,\frac{\lambda}{2} \qquad (n = 0, 1, 2, \ldots) \qquad (3)$$

In practice, an antinode never occurs quite at the end of an open pipe. Its position is just beyond the end of the pipe, the maximum displacement slightly overshooting the end (see figs. (a) and (b)). Thus for the first resonance ($n = 0$ in (3)) the length of the tube will be almost a quarter of wavelength, and $\lambda/4 = L + E$, where L is the length of the air column and E is the end correction. Similarly, for the second resonance, $3\lambda/4 = L + E$, where L' is the length of the air column when the second resonance occurs. Thus, using our data,

$$\frac{\lambda}{2} = L' - L = (66.4 - 21.9) \text{ cm}$$

$$= 44.5 \text{ cm} \qquad\qquad \text{or } \lambda = 89 \text{ cm.}$$

Therefore the velocity of sound is

$$v = f\lambda = 384 \text{ s}^{-1} \times 89 \text{ cm} = 34176 \text{ cm} \cdot \text{s}^{-1}$$

$$= 341.8 \text{ m} \cdot \text{s}^{-1}$$

Further, E = l/4 – L = (22.25 – 21.9) cm = 0.35 cm.

The velocity of sound in a gas is proportional to the square root of the absolute temperature. Hence

$$V_0/V = \sqrt{T_0/T} = \sqrt{273^\circ K/286^\circ K} = 0.977.$$

$$\therefore\ v_0 = 0.977 \times 341.8 \text{ m} \cdot \text{s}^{-1} = 333.9 \text{ m} \cdot \text{s}^{-1}.$$

Problem 17. When a Kundt's tube contains air, the distance between several nodes is 25 cm. When the air is pumped out and replaced by a gas, the distance between the same number of nodes is 35

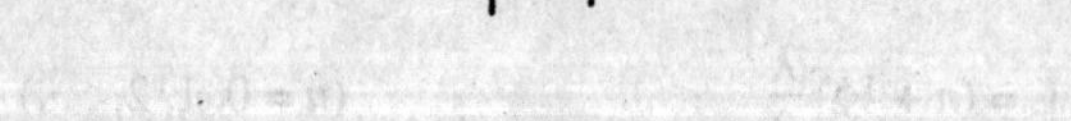

Fig.

cm. The velocity of sound in air is 340 m · s^{-1}. What is the velocity of sound in the gas?

Solution: A Kundt's tube (see the figure) is a glass tube with sawdust spread over its interior. A piston P is set into longitudinal vibration and the wall W moved until a standing wave pattern of nodes and antinodes is set up within the tube. The sawdust will accumulate at all the nodal points, for these are the points of the wave for which there is no vibration of the gas. By measuring the distance d between the nodes, we may find the velocity of sound of the gas in the tube. Let the number of nodes involved be $2n + 1$. There are $2n$ intervals between these $2n + 1$ nodes, and the total distance thus corresponds to $2n$ half-wavelengths, or to n full wavelengths.

The frequency of the emitted sound is the same in both cases, for it depends only on the frequency of the vibrating membrane which is causing the standing wave within the Kundt's tube. Thus $c_A = f\lambda_A$ and $c_G = f\lambda_G$.

Since there are, in the case of the air filled tube, n wavelengths in 25 cm of space, then the number of centimetres of space one wavelength will contain is found by the following proportion.

$$\frac{25\text{cm}}{\text{n wavelength}} = \frac{\text{x cm}}{\text{1 wavelength}} = \lambda_A$$

Similarly for the case of the gas filled tube,

$$\frac{35\text{ cm}}{\text{n wavelength}} = \frac{\text{x cm}}{\text{1 wavelength}} = \lambda_G$$

Thus

$$\frac{c_G}{c_A} = \frac{\lambda_G}{\lambda_A} = \frac{(35/n)\ cm}{(25/n\ cm} = \frac{7}{5}\ .\ \therefore\ c_G = \frac{7}{5} \times 340\ \text{m . s}^{-1}$$
$$= 476\ \text{m . s}^{-1}$$

If the number of nodes involved is 2n, then there are 2n –1 intervals between these 2 nodes. The total distance thus corresponds to 2n–1 half-wavelengths, or to $n-\frac{1}{2}$ full wavelengths. The same analysis as before holds and

$$\frac{c_G}{c_A} = \frac{\lambda_G}{\lambda_A} = \frac{\left(35/n - \frac{1}{2}\right) cm}{\left(25/n - \frac{1}{2}\right) cm} = \frac{7}{5}$$

The same result is obtained, as it must, if an odd number of nodes are counted as if an even number of nodes are counted.

Problem 18. (a) What will be the frequencies of the first and the second overtones of a pipe closed at one end of length 2 ft? (b) What will be the frequencies of the first and second overtones of an open pipe 2.5 ft long? (c) Will there be any common beat frequency between these overtones?

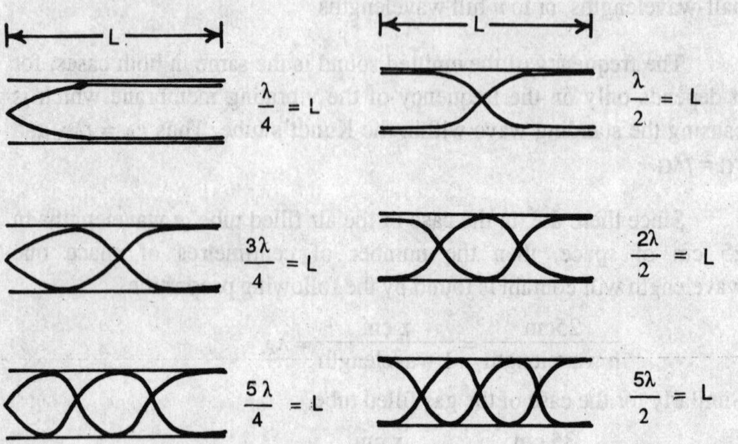

Fig.

Solution: (a) Not all wave shapes can be fitted into a closed (or semi-closed) pipe. Only those waves which satisfy certain boundary conditions can exist in the enclosure. Only those waves which have a node (a point of zero amplitude) at a closed end of the pipe and which have an antinode (a point of maximum amplitude) at an open end, can exist in the pipe. The wave with the lowest frequency which can exist in a pipe closed at one end is one which has only one node (at the closed end) and one antinode (at the open end). (See figure A). Its wavelength is $\lambda = 4\lambda$ and its frequency n_f (the fundamental frequency) is, using the relation $\lambda n_f = v$, where v is the velocity of sound,

$$n_f = \frac{v}{\lambda} = \frac{v}{4l} = \frac{1100\,ft/sec}{4 \times 2\,ft} = 137.5\ \text{sec}^{-1}$$

The first overtone contains two nodes and two antinodes. The second overtone contains three nodes and three antinodes. In general, the nth overtone will contain $n + 1$ nodes and $n + 1$ antinodes. The first and second overtones of the pipe will then be the third and the fifth harmonic respectively, i.e., frequencies 3 times and 5 times that of the fundamental (see figure A).

$$\therefore\ n_{1st} = 3\ (1367.5) = 412.5\ \text{sec}^{-1}$$

$$n_{2nd} = 5\ (137.5) = 687.5\ \text{sec}^{-1}$$

The only waves which can exist in an open pipe are those which have antinodes at both ends, of the pipe. The enclosed wave with the lowest possible frequency, n_f (the fundamental) space will contain two antinodes and one node (see figure B). The first overtone contains three antinodes and two nodes. The second overtone contains four antinodes and three nodes. In general, the nth overtone contains $n + 2$ antinodes and $n + 1$ nodes. The first and second overtones will then be the second and third harmonics, i.e., frequencies 2 times and 3 times that of the fundamental.

The wavelength of the fundamental is $\lambda = 2l$ and its frequency n_f is

$$n_f = \frac{v}{\lambda} = \frac{v}{2l} = \frac{1100\ ft/sec}{2 \times 2.5\ ft} = 220\ \text{sec}^{-1}$$

Hence

$$n'_{1st} = 2\ (220) = 440\ \text{sec}^{-1}$$

$$n'_{2nd} = 3\ (220) = 660\ \text{sec}^{-1}$$

The frequency difference for the first overtones is 440 – 412.5 = 27.5. The frequency difference for the second overtones is 687.5 – 660 = 27.5.

Thus a common beat frequency is 27.5 cycles per sec.

Problem 19. A source emits sound waves at 1000 cycles/second (frequency f_0) and the speed of the wave with respect to the source is 1000 ft/second (v_w).

(a) Calculate the wavelength a listener observes if he is at rest with respect to the source (λ_0)?

(b) If the source moves with a velocity of 100 ft/second (v_s)

towards the listener, what frequency (f_{sl}) and wavelength (λ_{sl}) does he observe?

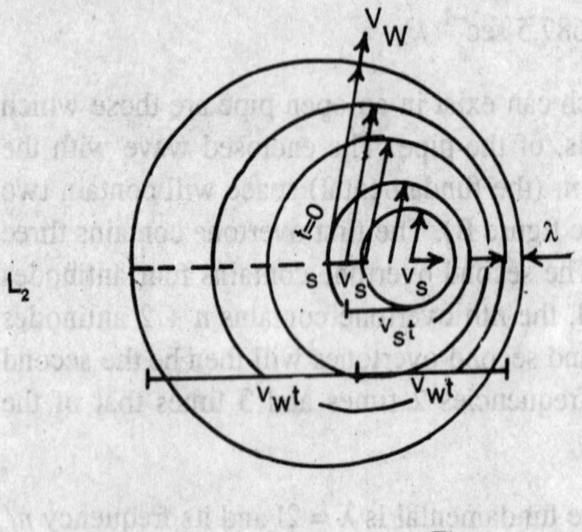

Fig.

(c) If the source moves at a velocity vs away from the listener, what frequency (f_{s2}) and wavelength (λ_{s2}) does he observe?

Solution: Everyone has heard the drop in pitch of a passing train's whistle. The Doppler effect is the apparent change of wave frequency observed by a listener when he and the wave source are in relative motion. In this problem, we deal with sound waves and derive the change in observed frequency and wavelength for two kinds of relative motion: motion of the observer and motion of the source. For any observer $v = \lambda f$, which means that the velocity at which he observes the wave to move is equal to the product of its observed frequency and wavelength. This is no more than an application of the conventional velocity definition $v = \Delta s/\Delta t$ if we remember $f = 1/T$ where T is the wave's period. Then $v = \lambda f = \lambda/T$ where λ is the observed displacement in time T.

(a) Using $v = \lambda f$, $v_w = \lambda_0 f_0$, $= \lambda_0 = v_w/f_0 = \dfrac{1000 \text{ ft/second}}{1000/\text{second}} = 1$ ft.

$$\frac{900 \text{ ft/sec}}{1000/\text{sec}} = 0.9 \text{ ft.}$$

(b) It is known that the velocity of waves in a medium depend only on the mechanical properties of that medium. When the source moves towards the listener (see L_1 in figure) a "bunching up" of waves is noted by the listener, as shown in the figure. We seek a mathematical relation to determine by what amount their wavelength is effectively reduced, and how the frequency is affected.

In a time, *t*, the source emits $t/T = f_0t$ waves. Along the line between listener and source these waves are spread over a distance $v_wt - v_st$. Thus, if the wavelength is smaller by a constant amount, it must now be

$$\lambda_{s1} = \frac{v_wt - v_st}{f_0t} = \frac{v_w - v_s}{f_0} = \frac{1000\,ft/sec - 100\,ft/sec}{1000/sec}$$

$$= \frac{900\,ft/sec}{1000/sec} = 0.9 \text{ ft.}$$

As the speed of the wave is unchanged, the listener will observe an increase in the wave's frequency, according to

$$v_w = \lambda_{s1} f_{s1};\ f_{s1} = v_w/\lambda_{s1} = \frac{1000\,ft/sec}{.9ft} = 1111 \text{ sec}^{-1}$$

(c) As the source recedes a listener observes the waves to occupy more space than those of a stationary source, as can be seen by looking at L_2 in the figure. In a time, *t*, f_0t waves are emitted, spread over a distance $v_wt + v_st$. As before, we find the wavelength

$$\lambda_{s2} = \frac{v_wt + v_st}{f_0t} = \frac{v_w + v_s}{f_0} = \frac{1000\,ft/sec + 100\,ft/sec}{1000/sec}$$

$$= 1.1 \text{ ft}$$

And, as in (b), we have for the observed frequency,

$$f_{S2} = \frac{v_W}{l_{S2}} = \frac{1000\,ft/sec}{1.1\,ft} = \frac{909}{sec}.$$

Problem 20. A researcher notices that the frequency of a note emitted by an automobile horn appears to drop from 284 cycles . s^{-1} to 266 cycles . s^{-1} as the automobile passes him. From this observation he is able to calculate the speed of the car, knowing that the speed of sound in air is 1100 ft . s^{-1}. What value does he obtain for the speed?

Solution: This is an example illustrating the Doppler effect. When there is no movement of the surrounding medium the relation between the frequency as heard by a moving observer and that emitted by a moving source is

$$\frac{f_L}{u \pm v_L} + \frac{f_s}{u + v_s}$$

which f_L is the frequency heard by the listener, f_s the frequency emitted by the moving source, v_L the velocity of the listener, v_s the velocity of the source, and u the velocity of sound $\left(= 1100\, ft \,.\, s^{-1}\right)$. The upper signs (+ left side of equation, – right side) correspond to the source and observer moving along the line joining the two and approaching, each other and the lower signs (– left, + right) correspond to source and observer receding from one another.

In this case the frequencies heard by the stationary listener (v_L = 0) will be $f_L = uf_s/\,(u + v_s)$. As the automobile approaches the observer he records a frequency of 284 cycles . s^{-1}, and as the automobile moves away from him, he records 266 cycles . s^{-1}. Thus

$$284\ s^{-1} = \frac{uf_s}{u - v_s} \qquad (1)$$

and

$$266\ s–1 = \frac{uf_s}{u + v_s} \qquad (2)$$

Dividing (1) and (2)

$$\frac{u + v_s}{u - v_s} = \frac{284}{266}$$

$$266\ (u + vs) = 284\ (u - vs)$$

$$(266 + 284)\ v^s = (284 - 266)u$$

or

$$\frac{v_s}{u} = \frac{18}{550}$$

$$\therefore\ v_s = \frac{18}{550} \times 1100\ \text{ft} \cdot s^{-1} = 36\ \text{ft} \cdot s^{-1} = 36\ \text{ft} \cdot s^{-1}$$

$$= 36\ \text{ft} \,.\, s^{-1} \times \frac{1\ mile}{5280\, ft} \times \frac{60s}{1min} \times \frac{60min}{1hr} = 24.5\ \text{mph}$$

Problem 21. An airplane is flying at Mach 0.5 and carries a sound source that emits a 1000-Hz signal. What frequency sound does a listener hear if he is in the path of the airplane after the airplane has passed?

Solution: The frequency of a wave disturbance depends on the relative motion of the source and observer. This phenomenon is called the Doppler effect and can be determined. The wavelength λ of a wave can be defined as

$$\lambda = \frac{\text{distance}}{\text{no. of waves}}$$

The no. of waves that a source of frequency v_s emits in time t is just v_{st}. If the medium permits the waves to travel at velocity v then the distance they cover in time t due to their own motion is v_t. Since the source is also moving toward the listener at velocity v_s and covers a distance $v_s t$, this means that the waves have a distance $v_t - v_s t$ to be spread out in. Therefore

$$\lambda = \frac{vt - v_s t}{\upsilon_s t} = \frac{v - v_s}{\upsilon_s}$$

The frequency v_L of the wave as observed by the listener as the source moves toward him is

$$\upsilon_L = \frac{v}{\lambda_L} = v \times \frac{\upsilon_s}{v - v_s} = \upsilon_s \left(\frac{v}{v - v_s}\right) = \frac{\upsilon_s}{1 - v_s/v}$$

If the source moves away from the listener, the v_s is considered negative in the above expression.

A speed of Mach 0.5 means that the airplane moves at half the speed of sound in air. Therefore $v_s/v = 0.5$. As the airplane moves toward the listener

$$\upsilon_L = \frac{\upsilon_s}{1 - 0.5} = 2\upsilon s$$

so that the listener hears a 2000-Hz sound.

After the airplane has passed the listener, the frequency he hears is

$$\upsilon_L = \frac{\upsilon_s}{1 + v_s/v}$$

$$= \frac{v_s}{1 + 0.5} = \frac{2}{3}\upsilon_s$$

so that the listener hears a 667 - Hz sound.

4

Doppler Effect

Problem 1. A laser emits a monochromatic light beam, of wavelength λ, which falls normally on a mirror moving at velocity v. What is the beat frequency between the incident and reflected light?

Solution: When a source of light is in motion relative to an observer or vice versa, the light waves exhibit a change in frequency as seen by that observer. This phenomenon is known as the Doppler effect, and the Doppler shift for light is given as

$$\upsilon_{observed} = \upsilon\sqrt{\frac{c \pm v}{c \mp v}} \qquad (1)$$

where v is the frequency of the light in an inertial frame is reference, v is the speed of the source relative to the observer, and c is the speed of light. In this problem v is the speed of the mirror. Let us choose the mirror to be moving away from the laser, so that it absorbs the light as if it were an observer moving away at speed v; it then re-emits the light as if it were a source moving away at speed v. In both cases the source and observer are separating from one another, so the frequency decreases. To get $\upsilon_{observed} < \upsilon$, we need the lower signs in equation (1).

Now, when the mirror absorbs the light, we have

$$\upsilon_{observed} = \upsilon\sqrt{\frac{c - v}{c + v}} \qquad (2)$$

Similarly, when it re-emits the light, the frequency, shift is given by

$$\upsilon_{emitted} - \upsilon_{absorbed} \sqrt{\frac{c \pm v}{c + v}} = \upsilon\left(\frac{c - v}{c + v}\right) \qquad (3)$$

The beat frequency is the difference in frequencies between the incident and reflected light:

$$\upsilon_{beat} = \upsilon - \upsilon_{emitted}$$

$$\upsilon_{beat} = \upsilon - \upsilon\left(\frac{c-v}{c+v}\right)$$

$$= \upsilon\left[1 - \frac{c-v}{c+v}\right]$$

$$= \upsilon\left[\frac{2v}{c+v}\right]$$

$$= \upsilon_{beat} = \frac{c}{\lambda}\left[\frac{2v}{c+v}\right]$$

when $v << c$, then $c + v = c$. In this case $\nu_{beat} = \frac{2v}{\lambda}$.

Problem 2. The Milky Way galaxy rotates once in 200 million years, and our sun is located about 30,000 light years from the galactic centre. As a result, the earth is moving through space relative to the oter galaxies. What is the observed Doppler shift in Å of the hydrogen line of 6563 Å for light coming from other galaxies? Consider two cases: (a) The line of observation is in the direction of the earth's motion: (b) The line of observation is perpendicular to the direction of the earth's motion; (b) Ignore other causes of oberved Doppler effects.

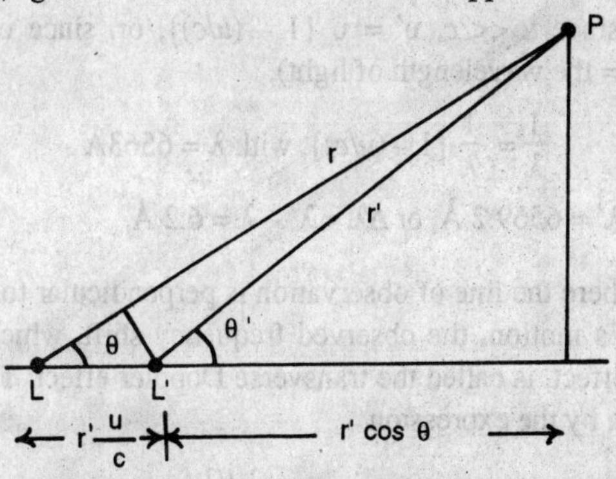

Fig.

Solution: The relativisitc Doppler shift is given by the expression

$$\upsilon' = \upsilon\sqrt{\frac{1 - u/c}{1 + u/c}}$$

Multiplying both the numerator and denominator by the factor $\sqrt{1 - u/c}$, one obtains the expression

$$\upsilon' = \upsilon\frac{1 - u/c}{\sqrt{1 - (u/c)^2}}$$

where υ' is the observed frequency of light of frequency υ in the frame of reference of the source; u is the relative source-observer velocity, and c is the speed of light.

For case (a), the light is travelling parallel to the earth's motion, and u is given by $u = R\Omega$, with R = 30,000 light-years, and

$$\Omega = \frac{2\pi}{200(10^6)} \text{ rad/year.}$$

Thus,

$$u = \frac{30.000(2\pi)}{200(10^6)} \text{ light-years (rad/year)}$$

$$= 9.425\ (10^{-4})\ c \text{ (where } c \text{ represents the speed of light)}$$

or

$$\left(\frac{u}{c}\right) = 9.425\ (10^{-4})$$

Now since $u << c$, $\upsilon' = \upsilon\ \{1 - (u/c)\}$, or, since $\upsilon' = c/\lambda'$ and $\upsilon = c/\lambda$ (λ = the wavelength of light),

$$\frac{1}{\lambda'} = \frac{1}{\lambda}\{1 - (u/c)\}, \text{ with } \lambda = 6563\text{Å}.$$

Thus $\lambda' = 6569.2$ Å, or $\Delta\lambda = \lambda' - \lambda = 6.2$ Å

(b) Where the line of observation is perpendicular to the direction of the earth's motion, the observed frequency shift, which is purely a relativistic effect, is called the transverse Doppler effect. The frequency shift is given by the expression

$$\upsilon = \upsilon'\ [1 - (u/sc)^2]^{1/2}.$$

The derivation of this effect is indicated in the figure. The generalized Doppler shift is given by

$$\upsilon + \upsilon' \frac{1 = (u/c)\cos\theta'}{\sqrt{1-(u/c)^2}}$$

where, as indicated, the observer (on the earth) is located at point P at rest in the unprimed coordinate system, and the source of light is at rest in the primed system. For transverse observation

$$\theta = \frac{\pi}{2},$$

$$\cos\theta' = -\frac{u}{c}$$

Thus, $$\upsilon = \upsilon' \frac{1-(u/c)^2}{\sqrt{1-(u/c)^2}} = \upsilon' \, [1-(u/c)^2]^{1/2}$$

or

$$\frac{1}{\lambda} = \frac{1}{\lambda'} [1-(u/c)^2]^{1/2}, \text{ with } (u/c) = 9.425\,(10^{-4})$$

Thus, $\lambda = 6563.003$ Å, or $\Delta\lambda = 0.003$Å.

5

Geometrical Optics

Problem 1. Prove that the vertical image observed in a plane mirror is the same distance behind the mirror as the object is in front of the mirror.

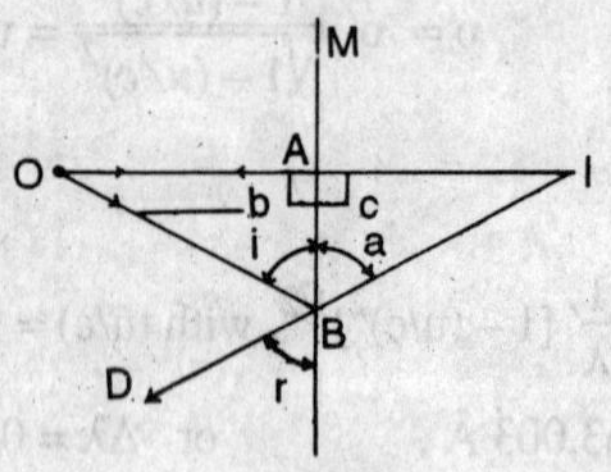

Fig.

Solution: As shown in the figure, let OA be the ray of light that strikes normal to the reflecting surface, while OB represents the ray that strikes the mirror at point B. The law of reflection states that the angle of incidence i equals the angle of reflection r,

$$i = r.$$

Rays DB and OA are extended back through the mirror to form triangle AIB, where point I is the apparent position of the image. Angle r must equal a; therefore, we obtain

$$a = r = i.$$

Angles b an c are equal since they are both right angles. Therefore, we have shown that the two triangles OAB and IAB are congruent, that is, coincide perfectly when superimposed because they share a common side, AB. We conclude that OA = IA and that the virtual image appears as far behind the mirror as the object is in front of the mirror.

Problem 2. What is the minimum length L or a wall mirror so that a person of height *h* can view herself from head to shoes?

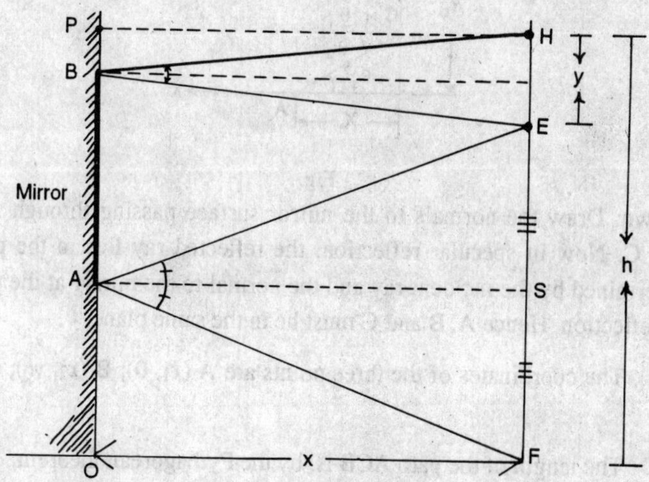

Solution: This is not easily solved by a diagram. We suppose that the person stands a distance x from the wall and that her eyes (E) are a distance y from the top of her head (H). To look at her toes (F) she looks at point A which is the point of reflection of a light ray from her foot. A must be at a height halfway between her eyes and feet (so that the angle of incidence equals the angle of reflection). Similarly to look at the top of her head she looks at point B. If OP = h and BP = $y/2$, then the length of the mirror is AB = PO – BP – OA = $h - y/2 - 1/2\,(h - y) = h/2$. Thus the minimum length of the mirror is $h/2$, and this does not depend on the distance x that the person is standing away from the mirror.

Problem 3. Prove that when light goes from one point to another via a plane mirror, the path chosen is the one which takes the least time.

Solution: Let the points be A and B, and let C be any general point on the mirror. Orient the diagram so that the x- and y-axes are as

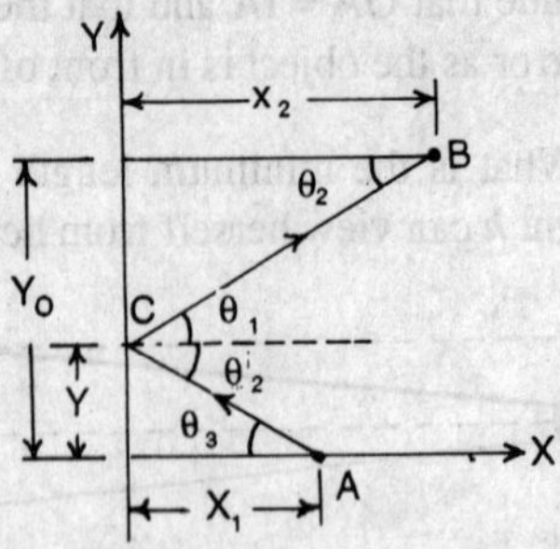

Fig.

shown. Draw the normals to the mirror surface passing through A, B and C. Now in specular reflection, the reflected ray lies in the plane determined by the incident ray and the normal to the mirror at the point of reflection. Hence A, B and C must be in the same plane.

The coordinates of the three points are A (x_1, 0); B (x_2, y_0); C (0, y).

The length of the path ACB is, by the Pythagorean theorem,

$$p = \sqrt{x_1^2 + y^2} + \sqrt{x_2^2 + (y_0 - y)^2}$$

But the time of travel of light by this path, the velocity of light being, c, is $t = p/c$. For the path to be travelled in minimum time, we must have $dt/dy = 0$, where y is the variable which changes with path. Thus

$$t = \frac{1}{c}\left[\left(x_1^2 + y^2\right)^{1/2} + \left(x_2^2 + (y_0 - y)^2\right)^{1/2}\right]$$

$$\frac{dt}{dy} = \frac{1}{c}$$

$$\left[\tfrac{1}{2}\,(x_1^2 + y^2)^{-1/2}\,(2y) + \tfrac{1}{2}\left(x_2^2 + (y_0 - y)^2\right)^{-1/2} \times 2\,(y_0 - y)\,(-1)\right]$$

$$\frac{dt}{dy} = \frac{1}{c}\left(\frac{y}{\sqrt{x_1^2 + y^2}} - \frac{(y_0 - y)}{\sqrt{x_2^2 + (y_0 - y)^2}}\right)$$

To find the values of y which make t a minimum, we set dt/dy = 0. Hence,

$$0 = \frac{1}{c}\left(\frac{y}{\sqrt{x_1^2 + y^2}} - \frac{y_0 - y}{\sqrt{x_2^2 + (y_0 - y)^2}}\right).$$

But some $c \neq \infty$, the quantity in the braces must be zero. Therefore,

$$\frac{y}{\sqrt{x_1^2 + y^2}} = \frac{y_0 - y}{\sqrt{x_2^2 + (y_0 - y)^2}} \qquad (1)$$

$$\sin \theta_1 = \frac{y}{\sqrt{x_1^2 + y^2}} \qquad \sin \theta_2 = \frac{y_0 - y}{\sqrt{x_2^2 + (y_0 - y)^2}}$$

Hence, (1) becomes

$$\sin \theta_1 = \sin \theta_2$$

and consulting the diagram we see that,

$$\theta_1 = \theta_2$$

which is the law of reflection. Since light reflected specularly always satisfies this condition, the light ray follows the path which takes the least tme.

Problem 4. An object is 12 feet from a concave mirror whose focal length is 4 feet. Where will the image be found?

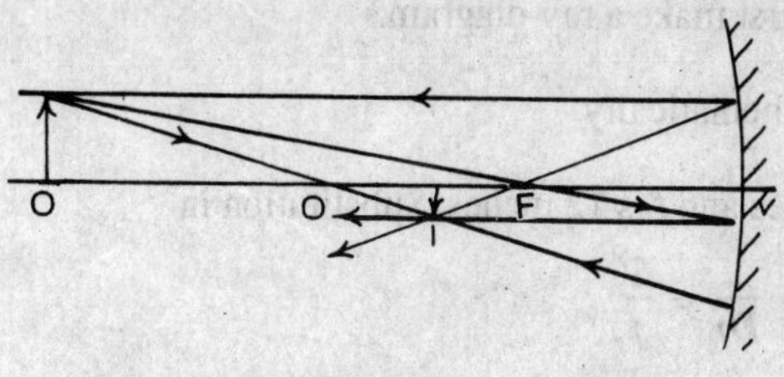

Solution: We first construct a ray diagram.

The image is a real image (inverted) between the centre of curvature and the focal point. It is smaller than the object.

We now solve mathematically.

Do is 12 feet and *f* is 4 feet. Substituting in

$$\frac{1}{D_O} + \frac{1}{D_I} = \frac{1}{f}$$

$$\frac{1}{12ft} + \frac{1}{D_I} = \frac{1}{4ft}$$

whence

$$D_I = 6\text{ft}$$

Problem 5. An object is 4 inches from a concave mirror whose focal length is 12 inches. Where will the image be formed?

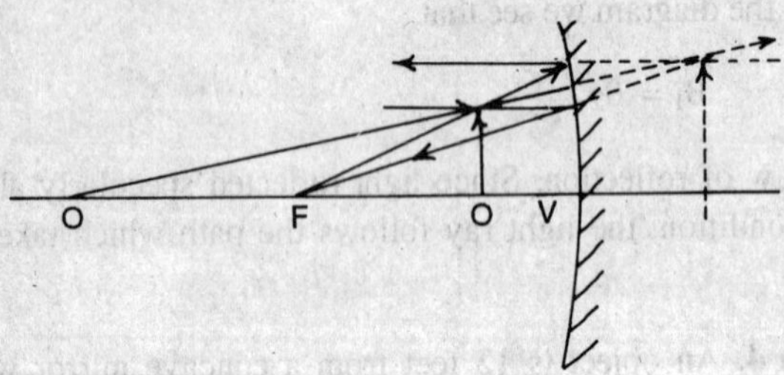

Fig.

Solution: First make a ray-diagram.

Solving mathematically

Do is 4 inches and *f* is 12 inches. Substitution in

$$\frac{1}{D_O} + \frac{1}{D_I} = \frac{1}{f}$$

$$\frac{1}{4in.} + \frac{1}{D_I} = \frac{1}{12in.}$$

whence $D_I = -6$ in.

which means that since D_I is negative the image is 6 inches from the mirror on the same side onthe mirror as the object and is virtual.

Problem 6. An object is placed 12 centimetres from a convex mirror whose focal length is 6 centimetres. Where will the image be formed?

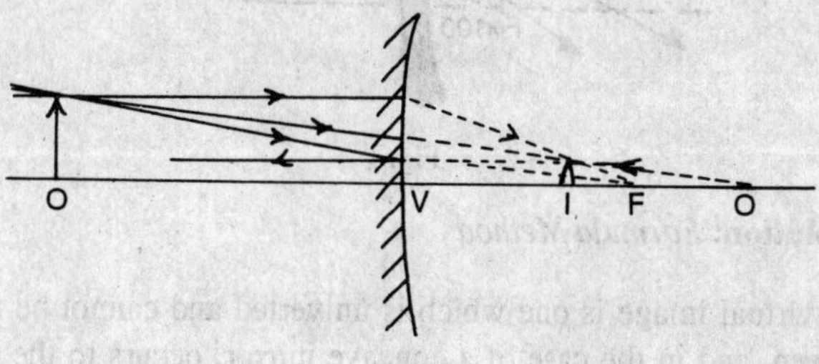

Fig.

Solution: First, we construct a ray-diagram.

Here we trace the path of two rays – one parallel to the axis, and one through the centre of curvature. The image is virtual and smaller than the object (for the convex mirror this is true for all positions of the object).

We may now solve mathematically.

D_O is 12 centimetres, f is –6 centimetres. Substituting in

$$\frac{1}{D_O} + \frac{1}{D_I} = \frac{1}{f}$$

$$\frac{1}{12cm} + \frac{1}{D_I} = \frac{1}{-6cm}$$

whence $D_I = -4\text{cm}.$

D_I is negative implies that the image is on the same side of the mirror as the object and is virtual.

Problem 7. Where would an object have to be located in front of a concave mirror in order to have a virtual image formed? Precisely where would it be located if the radius of the mirror were 20 cm and the image were 20 cm behind (i.e., to the right of) the mirror?

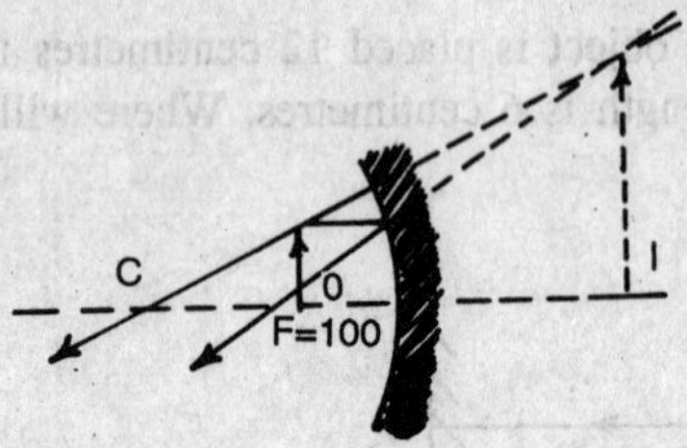

Fig.

Solution: *Formula Method*

A virtual image is one which is univerted and cannot be focused on a screen, and in the case of a concave mirror, occurs to the right of the mirror. Referring to the figure, we reason back to the intersection two light rays "emitted" from the tip of the image. The ray reflected back through the center of curvature must have originated there, since the angle of incidence is equal to the angle of reflection, and the angle of incidence is zero. The ray reflected through the focal point must have originated as a ray parallel to the axis. The intersection of these two rays is between the mirror and the focal point (see figure).

To find the exact position from the values specified in the problem, we use the mathematical relationship:

$$\frac{1}{0}+\frac{1}{i}=\frac{2}{R}$$

$$\frac{1}{0}-\frac{1}{20cm}=\frac{2}{20cm}$$

$$\frac{1}{0}=\frac{2}{20cm}$$

$$0=6\frac{2}{3}\text{ cm}$$

Note that i is negative because the image is located behind the mirror.

Ray-Diagram Method

To form a virtual image with a concave mirror, the object must lie between the focal point and the mirror.

For an image to be 20 cm to the right of the mirror, the ray approaching the head end of the image from the focal point must have been reflected through the focal point, where upon it must have been parallel to the axis before reflection.

The ray approaching the head end of the image from the centre of curvature must have experienced no change in direction.

Consequently, by tracing these two rays back to their origin, the head end of the object is located as in the diagram, at approximately 6 2/3 cm in front of the concave mirror. The diagram checks the calculations.

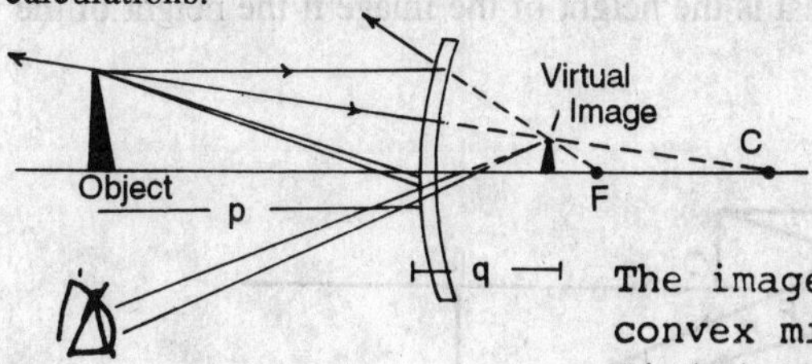

The image produced by a convex mirror is always diminished, erect, and vertical.

Fig.

Problem 8. A candle is held 3.0 in front of a convex mirror whose radius is 24 in. Where is the image of the candle?

Solution. The figure illustrates the conditions of the problem. The relative location of the candle can be found from the general mirror equation

$$\frac{1}{p}+\frac{1}{q}=\frac{1}{f}$$

where p is the distance between the candle and the mirror, q is the distance between the image and the mirror, and f is the focal length of the mirror. For a spherical mirror of small arc, as this is assumed to be, the focal length is approximately equal to one-half the radius of the mirror. Therefore $f = \frac{r}{2} = \frac{-24in}{2} = -12$ in.

$$\frac{1}{3.0in.} + \frac{1}{q} = \frac{-1}{12in}$$

$$\frac{1}{q} = \frac{-1-4}{12in.} = -\frac{5}{12in}$$

$$q = \frac{-12}{5} \text{ in.}$$

The negative sign for q indicates that the image lies behind the mirror and is a virtual image.

Problem 9. What type of mirror is required to form an image, on a wall 3 m from the mirror, of the filament of a headlight lamp 10 cm in front of the mirror? What is the height of the image if the height of the object in 5 mm?

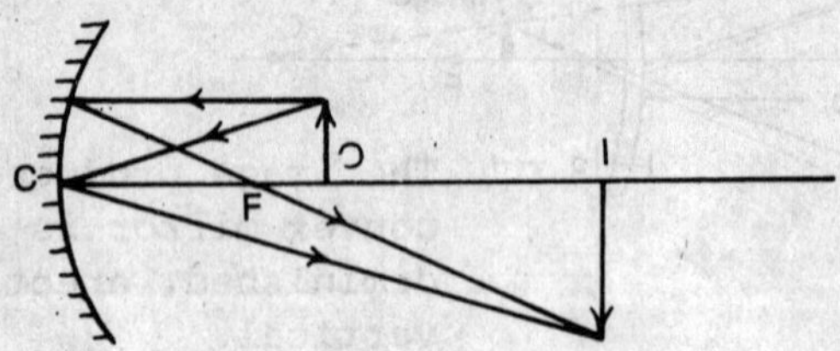

Fig.

Solution: The image I has to be real since it appears on the wall. Therefore, we must use a concave mirror and the object 0 must be placed beyond the focal point F for a real image (see the figure). We rule out a convex mirror since one cannot obtain images that can be shown on a screen with a convex mirror.

If p and q are the object and image distances from the mirror respectively, the first mirror equation is

$$\frac{1}{p} + \frac{1}{q} = \frac{1}{f}$$

where $f = \frac{R}{2}$ is the focal distance of the mirror. Therefore, we have

$p = 10$ cm, $\quad q = 300$ cm.

$$\frac{1}{10cm} + \frac{1}{300cm} = \frac{2}{R},$$

and

$$R = 19.4 \text{ cm.}$$

The radius R has a positive sign, as it is required for a concave mirror. The optical magnification m is

$$m = -\frac{q}{p}$$

where the negative sign is required because we want m to come out negative to signify an inverted image. The image height is obtained as follows.

$$h_1 = h_0 m = -h_0 \frac{q}{p} = -0.5 \text{ mm} \times \frac{300cm}{10cm} = -15 \text{ cm.}$$

The image is therefore 30 times bigger than the object and is inverted.

Problem 10. A small object lies 4 in. to the left of the vertex of a

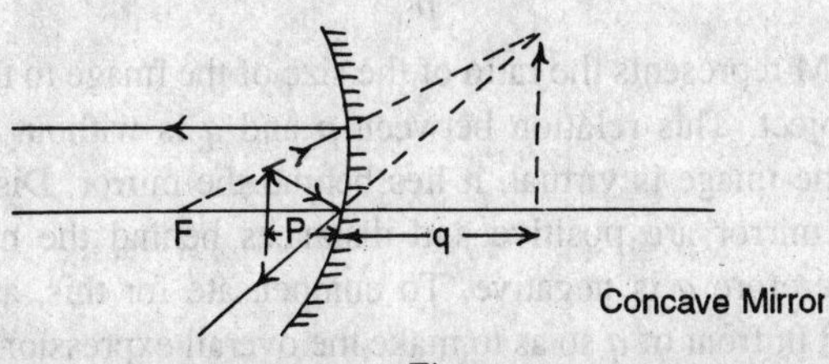

Fig.

concave mirror of radius of curvature 12 in. Find the position and magnification of the image.

Solution: The mirror equation for a concave mirror is

$$\frac{1}{p} + \frac{1}{q} = \frac{2}{R}$$

with a positive R. The object distance is $p = +4$ in. and R is +12 in. Then,

$$\frac{1}{4in} + \frac{1}{q} = \frac{2}{12in}$$

$$q = -12 \text{ in.},$$

Magnification m is given by

$$m = -\frac{q}{p}$$

$$= -\frac{-12in}{4in} = 3$$

The image is therefore 12 in. to the right of the vertex (q is negative), is virtual (q is negative), erect (m is positive), and 3 times the height of the object. See the figure.

Problem 11. A man has a concave shaving mirror whose focal length is 20 in. How far should the mirror be held from his face in order to give an image of two-fold magnification?

Solution: An erect, virtual, magnified image is desired. With q as the distance between the mirror and image, and p the distance between the mirror and the man's face, the equation.

$$M = \frac{q}{p}$$

can be used. M represents the ratio of the size of the image to the size of the actual object. This relation between p and q is without regard to sign. Since the image is virtual, it lies behind the mirror. Distances in front of the mirror are positive and distances behind the mirror are negative. Therefore q is negative. To compensate for this, a negative sign is placed in front of q so as to make the overall expression positive. For a two-fold magnification,

$$M = \frac{-q}{p} = 2$$

$$q = -2p.$$

Substitution in the general mirror equation

$$\frac{1}{p} + \frac{1}{q} = \frac{1}{f}$$

gives $$\frac{1}{p} + \frac{1}{-2p} = \frac{1}{20in.}$$

$$\frac{2-1}{2p} = \frac{1}{20in.}$$

$$p = 10 \text{ in.}$$

Problem 12. Derive the mirror formula for rays incident on a mirror of radius of curvature R, if the rays make a small angle with the mirror's axis.

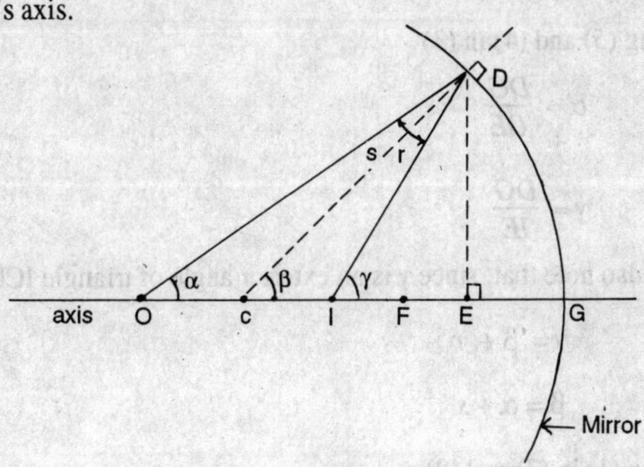

Fig.

Solution: Our proof will be totally geometrical. (See figure.) Ray OD is incident on mirror DG, at an angle s with the normal to the mirror CD. (C is the centre of curvature of the arc forming the mirror). It is reflected at an angle r, and intersects the mirror axis at point I. By the law of reflection, angle i is equal to angle r. Hence

$$s = r \tag{1}$$

Furthermore, $$\beta = \frac{DG}{CG} \tag{2}$$

But CG is the mirror radius R, whence

$$\beta = \frac{DG}{R}$$

Also $$\tan \alpha = \frac{DE}{OE}$$

and $$\tan \gamma = \frac{DE}{IE} \tag{3}$$

However, if α, β and γ are small angles (that is, ray OD is close to the mirror axis) we may write

$$\tan\alpha \approx \alpha \tag{4}$$

$$\tan\gamma \approx \gamma$$

Furthermore, $\quad DE \approx DG \quad (5)$

Using (5) and (4) in (3)

$$\alpha \approx \frac{DG}{OE}$$

$$\gamma \approx \frac{DG}{IE} \tag{6}$$

We also note that, since γ is an exterior angle of triangle ICD

$$\gamma = \beta + r \tag{7}$$

Similarly, $\quad \beta = \alpha + s \quad (8)$

Using (1) in (7) and (8)

$$\gamma = \beta + s \tag{9}$$

$$\beta = \alpha + s \tag{10}$$

Eliminating s in (9) and (10),

$$\gamma - \beta = \beta - \alpha$$

or $\quad 2\beta - \gamma - \alpha = 0 \quad (11)$

Substituting (6) and (2) in (11)

$$\frac{2DG}{R} - \frac{DG}{IE} - \frac{DG}{OE} = 0$$

or $\quad \dfrac{1}{IE} + \dfrac{1}{OE} = \dfrac{2}{R} \quad (12)$

However, if α, β and γ are small,

$$IE \approx IG = i$$

$$OE \approx OG = O \quad (13)$$

where i and O are the distances of the image from the mirror (image distance) and the object from the mirror (object distance). Hence, using (13) in (12) we have the mirror formula

$$\frac{1}{i} + \frac{1}{O} = \frac{2}{R}$$

Problem 13. How fast does light travel in glass of refractive index 1.5?

Solution: By definition, refractive index n is the ratio of the velocity of light in vacuum (3.00×10^{10} cm/sec) to the velocity of light in the medium in question.

Therefore $$n = \frac{3 \times 10^{10}\ cm/s}{v} = 1.5$$

Therefore $$v = \frac{3 \times 10^{10}\ cm/s}{1.5}$$

$$= 2.00 \times 10^{10}\ \text{cm/sec.}$$

Problem 14. An incident wavefront of light makes an angle of 60° with the surface of a pool of water. The speed of light in water is 2.3×10^8 m/s. What angle does the refracted wavefront make with the surface of water?

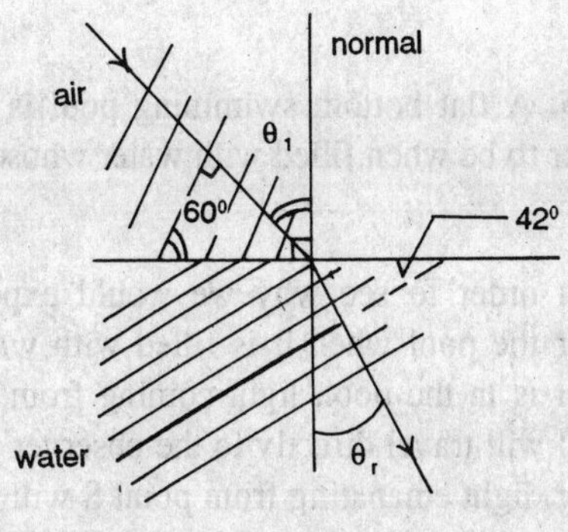

Fig.

Solution: The angle θ_1 between the incident ray and the normal to the surface (as shown in the figure), equals the angle between the incident wavefront and the water surface, and

$$\theta_i = 60^\circ$$

Snell's Law, relating the angle of incidence θ_i, to the angle of refraction, θ_r of the light, is

$$n_1 \sin \theta_i = n_2 \sin \theta_r$$

where n_1 and n_2 are the refractive indices of air and water, respectively. Hence

$$\sin \theta_r = \frac{n_1}{n_2} \sin \theta_i$$

But

$$n_1 = \frac{\text{speed of light (vacuum)}}{\text{speed of light (air)}} \qquad n_2 = \frac{\text{speed of light (vacuum)}}{\text{speed of light (water)}}$$

$$\text{Hence } \sin \theta_r = \left(\frac{2.3 \times 10^8 \ m/s}{3 \times 10^8 \ m/s} \right) \sin 60^\circ = .664$$

or $\qquad \theta_r = 42^\circ$

θ_r also equals the angle the refracted wavefront makes with the water surface.

Problem 15. A flat bottom swimming pool is 8 ft. deep. How deep does it appear to be when filled with water whose refractive index is 4/3?

Solution: In order to see why we would expect to observe a different depth for the pool when it is filled with water, examine the figure. If no water is in the pool, light coming from a point S on the bottom of the pool will travel directly to the observer's eye. If the pool is filled with water, light emanating from point S will be refracted at P, as shown. Upon reaching the observer's eye, the light appears to be coming from Q and he perceives the depth of the pool to be the distance OQ, rather than the actual depth OS. Our problem is to find the distance

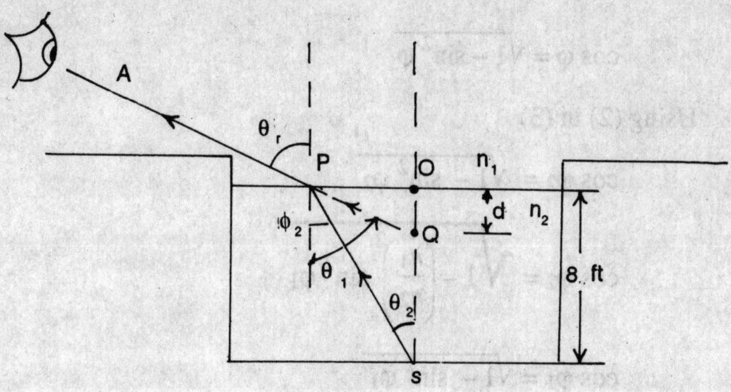

Fig.

d.

Note that, from the figure,

$$\tan \varphi_1 = \frac{OP}{d}$$

$$\tan \varphi_2 = \frac{OP}{8} \text{ ft.}$$

Hence

$$\frac{\tan \varphi_1}{\tan \varphi_2} = \frac{OP}{d} \cdot \frac{8\,ft.}{OP} = \frac{8\,ft.}{d}$$

and

$$d = \frac{(8\,ft)\tan \varphi_2}{\tan \varphi_1} \tag{1}$$

From Snell's Law,

$$n_1 \sin \varphi_1 = n_2 \sin \varphi_2$$

where n_1 and n_2 are the indices of refraction of air and water, respectively. Therefore

$$\left(\frac{n_1}{n_2}\right) \sin \varphi_1 = \sin \varphi_2 \tag{2}$$

To calculate the tangents in (1), we must also know $\cos \varphi_1$ and $\cos \varphi_2$. These we may find by observing that

$$\cos \varphi = \sqrt{1 - \sin^2 \varphi} \tag{3}$$

Using (2) in (3)

$$\cos \varphi_2 = \sqrt{1 - \sin^2 \varphi_2}$$

$$\cos \varphi_2 = \sqrt{1 - \left(\frac{n_1}{n_2}\right)^2 \sin^2 \varphi_1} \tag{4}$$

$$\cos \varphi_1 = \sqrt{1 - \sin^2 \varphi_1}$$

Hence

$$\tan \varphi_1 = \frac{\sin \varphi_1}{\cos\varphi_1} = \frac{\sin \varphi_1}{\sqrt{1 - \sin^2 \varphi_1}} \tag{5}$$

and using (2) with (4)

$$\tan \varphi_2 = \frac{\sin \varphi_2}{\cos \varphi_2} = \frac{\left(n_1/n_2\right) \sin \varphi_2}{\sqrt{1 - \left(n_1/n_2\right)^2 \sin^2\varphi_1}} \tag{6}$$

Substituting (5) and (6) in (1)

$$d = (8 \text{ ft}) \frac{\left(n_1/n_2\right) \sin \varphi_2}{\sqrt{1 - \left(n_1/n_2\right)^2 \sin^2\varphi_1}} \cdot \frac{\sqrt{1 - \sin^2 \varphi_1}}{\sin \varphi_1}$$

$$d = (8 \text{ ft})\,(n_1/n_2)$$

$$\sqrt{\frac{1 - \sin^2 \varphi_1]}{1 - \left(n_1/n_2\right)^2 \sin^2\varphi_1}} \tag{7}$$

Now, since we don't know the angle φ_1, we make an approximation. Suppose φ_1 is very small. (This means that the observer is looking almost directly down into the pool.) Then $\sin \varphi_1 \approx 0$ and the square root in (7) becomes 1. Therefore

$$d = (8 \text{ ft})\left(\frac{n_1}{n_2}\right) = (8 \text{ ft})\left(\frac{1}{4/3}\right) = 6 \text{ ft.}$$

The pool appears to be 6 ft. deep.

Problem 16. Show that the optical length of a light path, defined as the geometrical length times the refractive index of the medium in which the light is moving, is the equivalent distance which the light would have travelled in a vacuum.

Solution: Suppose that light travels a distance l in a medium of refractive index n. The optical length is then

optical length = nl

and, since $n = c/v$

optical length = cl/v

Here c and v are the speeds of light in vacuum and the medium, respectively.

But light travels with constant velocity in the medium, and hence $l/v = t$, where t is the time taken to traverse the light path.

$$nl = \frac{cl}{v} = ct = l_0$$

where l_0 is the distance the light would have travelled at velocity c, that is, in a vacuum. Thus the optical length is the equivalent distance which the light would have travelled in the same time in a vacuum.

Problem 17. Describe the phenomena of the critical angle in optics. What is the critical angle for a glass-air interface, if the index of refraction of glass with respect of air is 1.33?

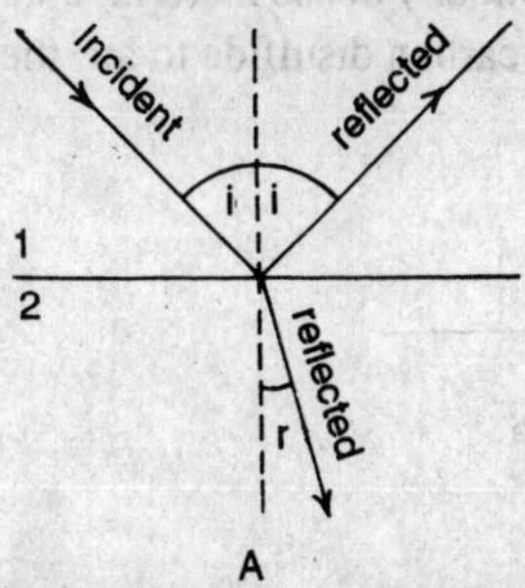

A

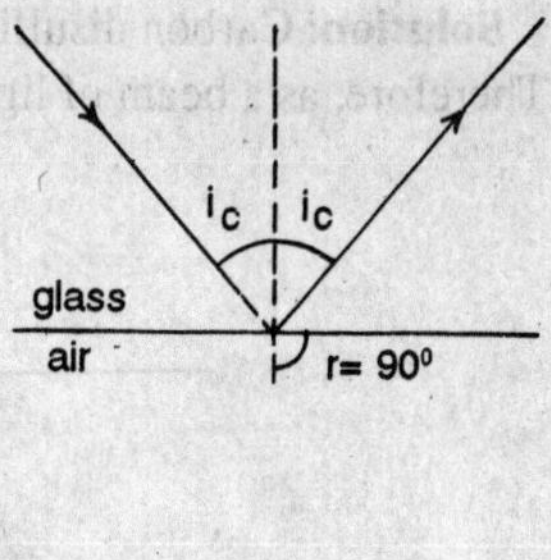

B

Solution: Consider two media, 1 and 2, such that the index of refraction of 1 with respect to 2 is less than unity, as shown in Fig. A.

That is, medium 1 is "denser", and the angle of refraction will be greater than the angle incidence.

In general part of the incident ray is reflected and part refracted. As the angle of incidence is increased, the angle of refraction will increase until $r = 90^\circ$. At this critical angle, i_c, (see Fig. B), we have

$$\frac{\sin i_c}{\sin 90^\circ} = \sin i_c = n_{21}$$

At the critical angle, and for values of i greater than ic, refraction cannot occur and all the energy of the incident beam appears in the reflected beam. This phenomenon is called total internal reflection.

The index of refraction of air with respect to glass is

$$\frac{1}{1.33} = 0.75$$

The critical angle for a glass-air interference is therefore

$$\sin i_c = 0.75$$

or $$i_c = 48.6^\circ$$

Thus, for angles of incidence ≥ 48,6°, total internal reflection will occur for a glass-air combination similar to the one shown in the figure.

Problem 18. What is the critical angle between carbon disulfide and air? (The index of refraction for Carbon Disulfide is 1.643).

Solution: Carbon disulfide is a more optically dense material than air. Therefore, as a beam of light passes from carbon disulfide to air, the

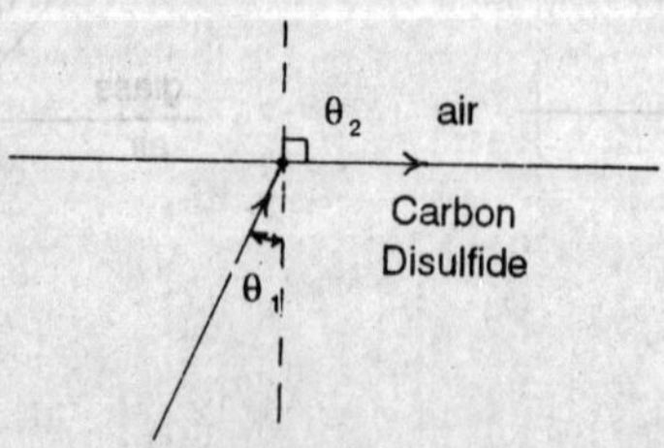

Fig.

angle of refraction is larger than the angle of incidence. There is an angle of incidence smaller than 90° for which the angle of refraction is equal to 90°, meaning that the beam of light emerges parallel to the boundary between the two mediums. This angle of incidence is called the critical angle. If the angle of incidence is greater than this value, the light will not escape from the carbon disulfide. It will be reflected back into the carbon disulfide following the regular law of reflection. Solving for the critical angle θ_1, let θ_2 be 90°. The index of refraction for carbon disulfide is 1.643 and for air it is 1.00 Using Snell's law

$$n_1 \sin\theta_1 = n_2 \sin\theta_2$$

$$1.643 \sin\theta_1 = 1.00 \sin 90^\circ$$

$$n\ \theta_1 = \frac{1.00}{1.643} = 0.648$$

$$\theta_1 = 37.4^\circ.$$

Problem 19. What is the critical angle of incidence for a ray of light passing from glass into water. Assume $n_{glass} = 1.50$ and $n_{water} = 1.33$.

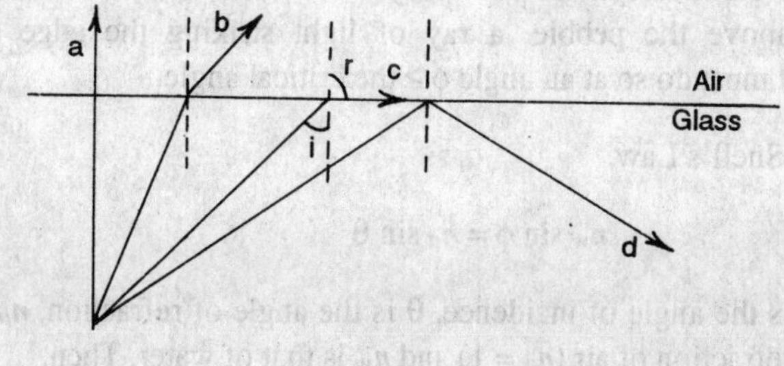

Fig.

Solution: When light passes from a more optically dense to a less optically dense medium, there is an angle of incidence *i*, called the critical angle, at which the light ray (ray *c* of the diagram) will be refracted parallel to the interface of the media. The angle of refraction *r* will be 90° in this case. Therefore, by Snell's Law, $n_1 \sin i = n_2 \sin r$. Also critical angle of incidence means $\sin r = 1$.

$$\therefore \sin i = \frac{n_2}{n_1} = \frac{1.33}{1.50} = .887$$

$\therefore\ i_c = 62^\circ$ (approx).

Problem 20. A small pebble lies at the bottom of a tank of water 24 feet deep. Determine the size of a piece of cardboard which, when floating on the surface of the water, directly above the pebble, totally obscures the latter from view.

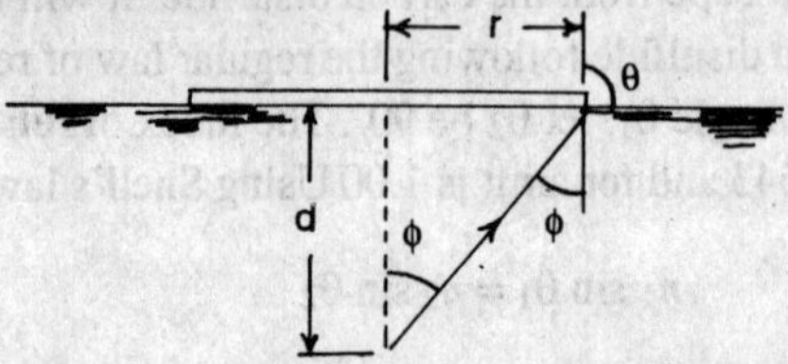

Fig.

Solution: We can consider the pebble as a point source of light. If the cardboard is big enough for its purpose, then all rays of light from the pebble which would be refracted into the air at the surface must be blocked off by the cardboard, and all rays striking the surface of the water outside the cardboard must be totally internally reflected (see figure).

The cardboard must obviously be circular, and, if its centre is directly above the pebble, a ray of light striking the edge of the cardboard must do so at an angle $\phi \geq$ the critical angle.

By Snell's Law,

$$n_w \sin\phi = n_a \sin\theta$$

where ϕ is the angle of incidence, θ is the angle of refraction, n_a is the index of refraction of air ($n_a = 1$) and n_w is that of water. Then,

$$n_w \sin\phi = 1$$

because $\theta = 90^\circ$ to satisfy the requirement that ϕ be a critical angle. Therefore, using the figure

$$n_w \frac{r}{\sqrt{r^2 + d^2}} = 1$$

$$r^2 + d^2 = n_w^2 r^2 \quad \text{or} \quad r^2 = \frac{d^2}{n_w^2 - 1}$$

$$r = \frac{24ft}{\sqrt{\frac{16}{9} - 1}} = \frac{72ft}{\sqrt{7}} = 27.2 \text{ ft.}$$

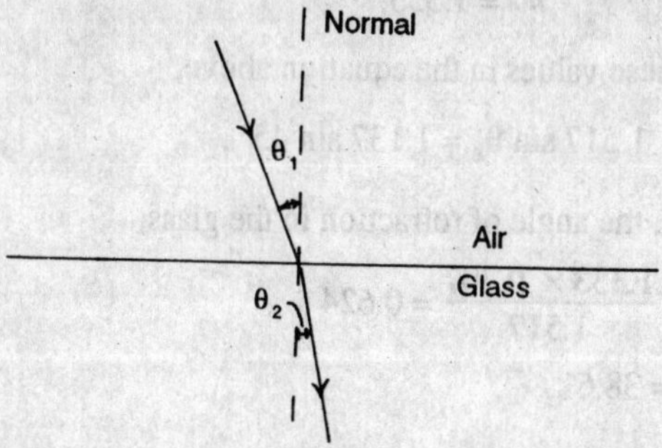

Fig.

Problem 21. A beam of light in air falls on a glass surface at an angle of incidence of 20.0°. Its angle of refraction in the glass is 12.4°. What is the index of refraction of the glass?

Solution: Since the beam of light passes obliquely from an optically less dense medium, air, to an optically denser medium, glass, the angle of incidence is larger than the angle of refraction. The equation for indices of refraction is

$$\frac{\sin \theta_1}{\sin \theta_2} = \frac{n_2}{n_1}.$$

Since the index of refraction for air is approximately one, this reduces to

$$n_2 = \frac{\sin \theta_1}{\sin \theta_2} = \frac{\sin 20.0^\circ}{\sin 12.4^\circ} = \frac{0.342}{0.215} = 1.58.$$

Problem 22. A ray of light in water is incident on a plate of crown glass at an angle of 45°. What is the angle of refraction for the ray in the glass?

Solution: The indices of refraction are inversely proportional to the sine of their respective angles, or, by Snell's law,

$$n_g \sin \theta g = n_w \sin \theta w.$$

From a table,

$$n_g = 1.517$$

$$n_w = 1.333.$$

Substituting these values in the equation above,

$$1.517 \sin \theta_g = 1.137 \sin 45^\circ .$$

Solving for θ_g, the angle of refraction in the glass,

$$\sin \theta_g = \frac{1.333 \times 0.707}{1.517} = 0.624$$

$$\theta_g = 38.6^\circ.$$

Notice that θg is less than θ_w. This occurs because the angle of incidence is greater than the angle of refraction when passing from an optically less dense medium, water, to an optically denser medium, glass.

Problem 23. A plate of glass 1.00 cm thick is placed over a dot on a sheet of paper. The dot appears 0.640 cm below the upper surface of the glass when viewed from above through a microscope. What is the index of refraction of the glass plate?

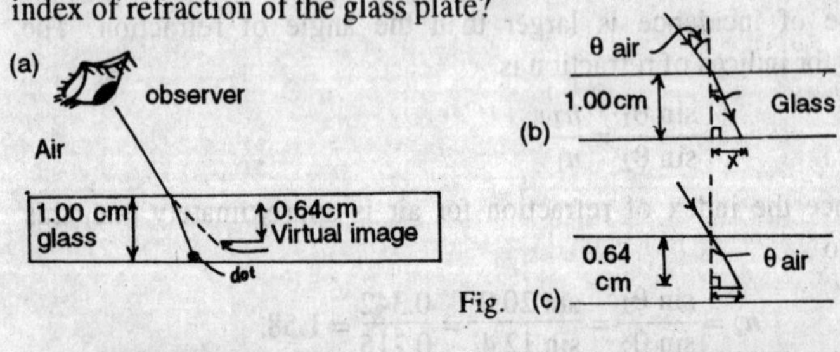

Fig.

Solution: The eye records the direction from which a light ray enters it as though it has travelled in a straight line from the source. It cannot compensate for refraction of the light. Therefore, the observer sees the dot 0.640 cm below the upper surface of the glass when in actually it is 1.00 cm below the surface.

Assuming the index of refraction of air to be one (n_{air} = 1.00029), we have from Snell's law,

$$n_{glass} \sin \theta_{glass} = n_{air} \sin \theta_{air} \quad (1)$$

$$n_g = \frac{n_g}{n_a} = \frac{\sin \theta_a}{\sin \theta_g} \text{ since } n_a \approx 1 \quad (2)$$

The sine function can be expressed as

$$\sin \theta = \tan \theta \cos \theta \quad (3)$$

We know from the figure that

$$\tan \theta_a = \frac{x}{0.64cm}$$

$$\tan \theta_g = \frac{x}{1.00cm}$$

From the trigonometric relation

$$\cos \theta = \sqrt{1 - \sin^2 \theta}$$

we have $\cos \theta_a = \sqrt{1 - \sin^2 \theta_a}$

$$\cos \theta_g = \sqrt{1 - \sin^2 \theta_g} \quad (4)$$

From equation (1),

$$\sin \theta_g = \frac{n_a}{n_g} \sin \theta_a \quad (5)$$

Substituting equation (5) into equation (4).

$$\cos \theta_g = \sqrt{1 - \left(\frac{n_a}{n_g}\right)^2 \sin^2 \theta_a}$$

Using the values found for the cosines and tangents of the angles and equation (3), substitute into equation (2)

$$n_g = \frac{\sin \theta_a}{\sin \theta_g} = \frac{\tan \theta_a \cos \theta_a}{\tan \theta_g \cos \theta_g}$$

$$\frac{\dfrac{x}{0.64cm} \sqrt{1 - \sin^2 \theta_a}}{\dfrac{x}{1.00cm} \sqrt{1 - \left(\dfrac{n_a}{n_g}\right)^2 \sin^2 \theta_a}}$$

The observer looks at the dot through a microscope. This means that he looks at the dot from directly above and θ_a must be very small. We can then make the approximation.

$$\sin \theta_a \approx \sin \theta^\circ = 0$$

Substituting this value in equation (6), we find

$$n_g = \frac{\frac{x}{0.64cm}\sqrt{1-0}}{\frac{x}{1.00cm}\sqrt{1-\left(\frac{n_a}{n_g}\right)^2 (0)}}$$

$$= \frac{1.00}{0.64} = 1.57$$

Problem 24. One end of a cylindrical glass rod is ground to a hemispherical surface of radius R = 2 cm. An object of height h_O = 1 mm. is placed on the axis of the rod, p = 8 cm to the left of the vertex. Find the image distance q and the image height h_I (a) when the rod is in air, (b) the rod is in water. The indices of refraction of glass and water are 1.50 and 1.33 respectively.

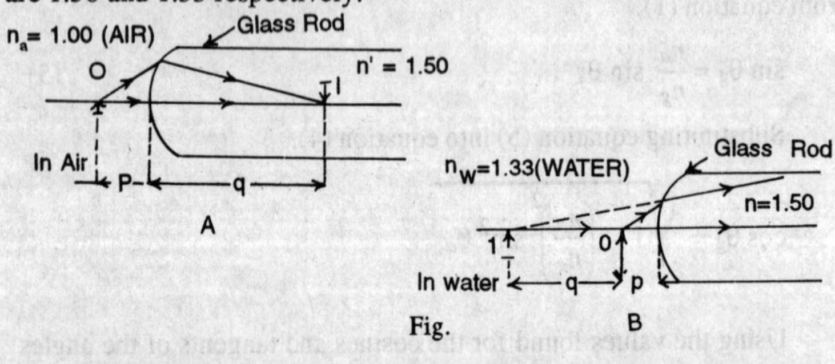

Fig.

Solution: (a) The optical equation for the spherical end of the rod (which is a spherical refracting surface), is given by

$$\frac{n}{p} + \frac{n'}{q} = \frac{n' - n}{R}$$

where the object is in the region of refraction index n, the rays are going from the region of refractive index n to the region of refractive index n', and R is the curvature of the refractive surface. The radius R for the glass hemisphere has a positive sign because the centre of curvature of the glass surface lies inside the glass, which is the real image region (see Fig. 1). R is negative when the curvature centre lies in the virtual image region.

As shown in Fig. 1 above, p = +8 cm, R = +2 cm. Therefore, the image distance q is obtained as

$$\frac{1.5}{q} + \frac{1}{8cm} = \frac{1.5-1}{23}cm$$

or

$$q = +12 \text{ cm}$$

q is positive, i.e., the image is formed at the right of the vertex of the glass surface and is real.

The magnification m for this problem is given by

$$m = -\frac{n}{n'}, \frac{q}{p}$$

$$= -\frac{1}{5.5} \times \frac{12cm}{8cm} = -1$$

that is, the image has the same height as the object, but is inverted,

$$h_I = -h_O = -0.1 \text{ cm}.$$

(b) When the rod is in water, the optical equation yields for q (Fig. 2)

or
$$\frac{1.5}{q} + \frac{1.33}{8cm} = \frac{1.50 - 1.33}{2cm}$$

$$q = -18 \text{ cm}.$$

Since q is negative, the image is formed in the water and is imaginary. For magnification, we have

$$m = -\frac{1.33}{1.5} \times \frac{-18cm}{8cm} = 1.99,$$

and the image is erect and has a height

$$h_I = 1.99\, h_O = 0.99 \text{ cm}.$$

Problem 25. A man standing symmetrically in front of a plane mirror with beveled edges can see three images of his eyes when he is 3ft from the mirror (see figure (a)). The mirror is silvered on the back, is 2ft 6 in. wide, and is made of glass of refractive index 1.54. What is the angle of bevel of the edges?

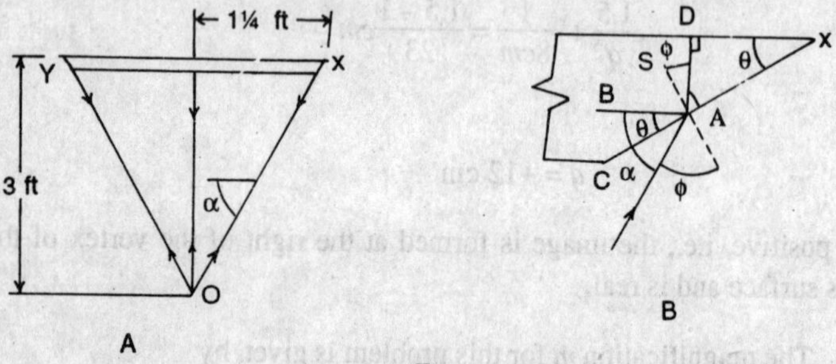

Fig.

Solution: The man can only see an image of his eyes if light leaves them, strikes the mirror, and is reflected back along the same path. The central image is thus formed by light traversing the perpendicular from his eyes to the mirror. The outer images are formed by light striking the beveled edges at the point A (see figure (b)) at an angle of incidence ϕ such that the angle of refraction ϕ' makes the refracted ray strike the silvered surface normally. This must be the case if the ray of light is to leave the beveled edge by the same path with which is arrived. The angle ϕ' lies between the normal to the beveled edge and the normal to the back surface. Since SAX = 90° (see figure (b)).

$$\phi' + \measuredangle DAX = 90^\circ$$

But $\measuredangle DAX = 90^\circ - \theta$

Hence $\phi' = 90^\circ - 90^\circ + \theta = \theta$

Draw BA, a construction line at A parallel to the back of the mirror. Angle BAC is also equal to θ.

But by Snell's Law $n_1 \sin \phi = n \sin \phi'$, where n_1 is the refractive index of air ($n_1 = 1$) and n is that of glass. Then $\sin \phi = n \sin f' = n \sin \theta$.

Also $a = \theta + (90 - \phi)$. (See figure (b)).

$$\sin[90 - (\alpha - \theta)] = n \sin \theta.$$

But $\sin(90 - \psi) = \cos \psi$ and

$$\cos(\alpha - \theta) = \text{n} \sin \text{q}.$$

By the trignometric relation for double angles

$$\cos \alpha \cos \theta + \sin \alpha \sin \theta = n \sin \theta.$$

$$\cos \alpha + \sin \alpha \tan \theta = n \tan \theta$$

$$\cos \alpha = \tan \theta \, [n - \sin \alpha]$$

$$\tan \theta = \frac{\cos\alpha}{n - \sin \alpha}$$

Looking at figure (a)

$$\cos \alpha = \frac{1\,1/4\,ft}{\sqrt{(1\,1/4\,ft)^2 + (3\,ft)^2}} = \frac{5}{13}$$

$$\sin \alpha = \frac{3\,ft}{\sqrt{(1\,1/4\,ft)^2 + (3\,ft)^2}} = \frac{12}{13}$$

whence $\tan \theta = \dfrac{5/13}{1.54 - (12/13)} = 0.625$

$$\theta = 32^\circ.$$

Problem 26. In hunting a submarine, two ships A and B, separated by 3000 ft, find a sonar echo on the line between them but 30° from vertical for A and 60° for B. However, 200 ft below the surface, the water temperature changes suddenly so that the velocity of sound in the depths is 0.9 that in the shallows. (a) For what depth will the depth charges be set, if the ship captains do not know of this? (b) What is the actual depth of the submarine? (c) What will the ship captains think the submarine's horizontal distance from ship A is? (d) What is the submarine's actual horizontal distance from ship A?

Solution: Figure (a) shows the actual paths of the sonar waves represented by solid lines, and their apparent paths as broken lines. The

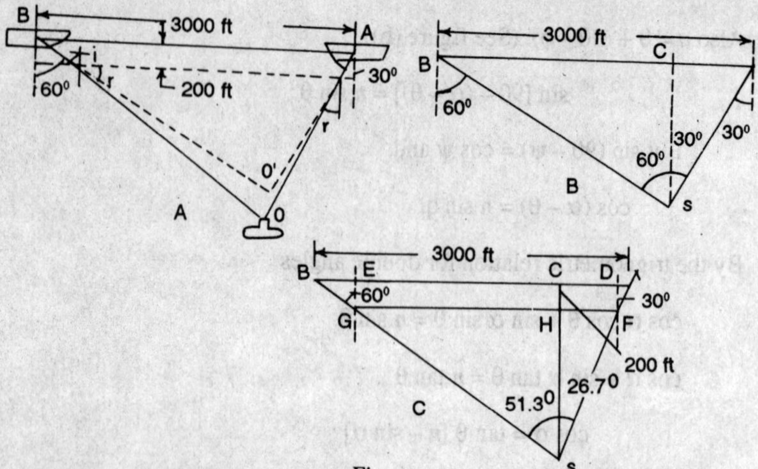

Fig.

submarine will appear to the ship captains to be at point 0′ when in reality it is at point 0, as shown.

$$AC + CB = 3000$$

$$\frac{AC}{CS} = \tan 30$$

$$\frac{CB}{CS} = \tan 60$$

$$AC + CB = CS\,(\tan 30 + \tan 60)$$

$$CS = \frac{AC + CB}{\tan 30 + \tan 60}$$

$$= \frac{3000\,ft}{0.5777 + 1.732}$$

$$= 1310 \text{ ft}$$

since $\tan 30 = 0.577$

$\tan 60 = 1.732$

(b) In actual fact the sound waves are refracted at a depth of 200 ft. By definition, the refraction is such that:

$$\frac{\sin i}{\sin r} = \frac{v_s}{v_d} = \frac{v_s}{0.9\, v_s} = \frac{1}{0.9}$$

where i is the angle that the path the sonar wave takes in the shallow water makes with an imaginary line normal to the interface between shallow and deep water. The letter r stands for a similar angle for the path in deep water, v_s and v_d represent the velocities of sound in the shallow and deep water, respectively.

If $i = 30°$, then $r = 26.7°$ and if $i = 60°$, $r = 51.3°$. We see from figure (c) that:

$$AD + DC' + C'E + EB = 3000 \text{ ft}$$

$$AD = 200 \tan 30$$

$$EB = 200 \tan 60$$

$$\frac{FH}{HS} = \tan 26.7 \qquad \frac{HG}{HS} = \tan 51.3$$

We can see from the diagram that:

$$DC' = FH$$

$$C'E = HG$$

Thus, from the first equation in the above group:

$$AD + FH + HG + EB = 3000 \text{ ft.}$$

$$200 \tan 30 + HS (\tan 26.7 + \tan 51.3) + 200 \tan 60 = 3000$$

$$HS = \frac{3000 - 115 - 346}{0.503 + 1.248}$$

$$= 1450 \text{ ft}$$

The actual depth of the submarine is 200 + 1450 = 1650 ft.

The horizontal distance of the submarine from ship A will appear to be AC in figure (b):

$$AC = 1310 \tan 30 = 755 \text{ ft}$$

The submarine's actual horizontal distance is AC' in figure (c):

$$AC' = 200 \tan 30 + 1450 \tan 26.7$$

$$= 115 + 730 = 845 \text{ ft.}$$

Thus without knowledge of the water temperature the depth charges will be shallow by 340 ft and off in a horizontal direction by 90 ft. That is, the explosion will be 350 ft away from the submarine, a near miss rather than a direct hit.

PRISMS

Problem 27. A ray of light enters the face BA of a right-angled prism of refracting material at grazing incidence. It emerges from the adjacent face AC at an angle θ to the normal. If ϕ_c is the critical angle for the material, show that $\sin \theta = \cot \phi_c$. (See figure).

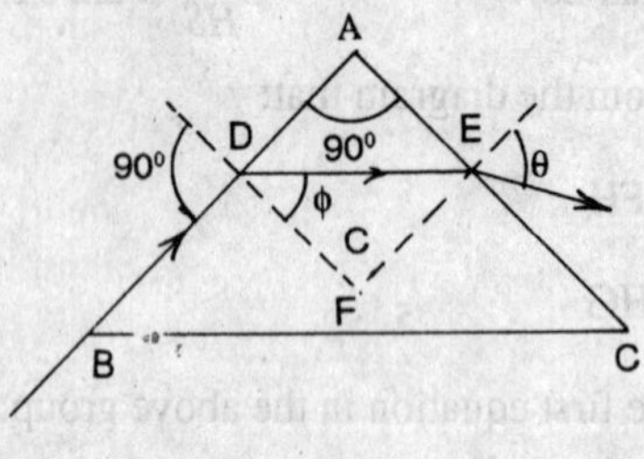

Fig.

Will a ray always emerge from AC? If not, explain what happens, and deduce for what values of the refractive index of the material the ray actually emerges.

Solution: Since the ray strikes the prism at grazing incidence, the refracted ray DE must enter at the critical angle ϕ_c. Now angles DAE, ADF, and AEF are all right angles. This follows because the prism is right-angled, and also since DF and FE are normals to the surfaces BA and AC. Thus DFE must be a right angle also, since the four angles of a quadrilateral (ADFE) add up to 360°.

The angles of a triangle must add up to 180° and therefore $\angle DEF = 90° - \phi_c$.

Applying Snell's law to the refraction at face AC, we have n_1 sin

$\angle DEF = n_2 \sin\theta$, where n_1 and n_2 are the refractive indices of the prism and air, respectively. Since the medium to the right of face AC is air, $n_2 = 1$, whence

$$n_1 \sin \angle DEF = \sin\theta \qquad (2)$$

At surface BA, we obtain

$$\cdot n_2 \sin 90^\circ = n_1 \sin\phi_c$$

or $\quad (1)\, n_1 \sin\phi_c$

Then $$\sin f\phi_c = 1/n \qquad (3)$$

Using (3) and (1) in (2)

$$\left(\frac{1}{\sin\phi_c}\right)(\sin(90 - \phi c)) = \sin\theta$$

But $\sin(90 - \phi_c) = \cos\phi_c$, whence

$$\frac{\cos\phi_c}{\sin\phi_c} = \cot\phi_c = \sin\theta$$

The ray will emerge from AC only if $\angle DEF$ is less than the critical angle. If it is greater, total internal reflection occurs and the ray is directed along AC.

Thus, for the ray to emerge, $\angle DEF < \phi_c$. That is, $90^\circ - \phi_c < \phi_c$ or $\phi_c > 45^\circ$. Using (3)

$$\therefore \sin\phi_c = \frac{1}{n_1} > \sin 45^\circ = \frac{1}{\sqrt{2}} \therefore n_1 < \sqrt{2}.$$

Problem 28. The easiest method of measuring the refracting angle of a prism is to direct a parallel beam of light on to the angle (vertex A in figure) and measure the angular separation of the beams reflected from the two sides of the prism containing the refracting angle. Show that this angular separation is twice the angle of the prism.

Solution: Consider three incoming rays, all parallel and striking the prism at points A, B and C. Erect normals to AB at A and B and to AC at A and C. Designate the anlges as in the diagram.

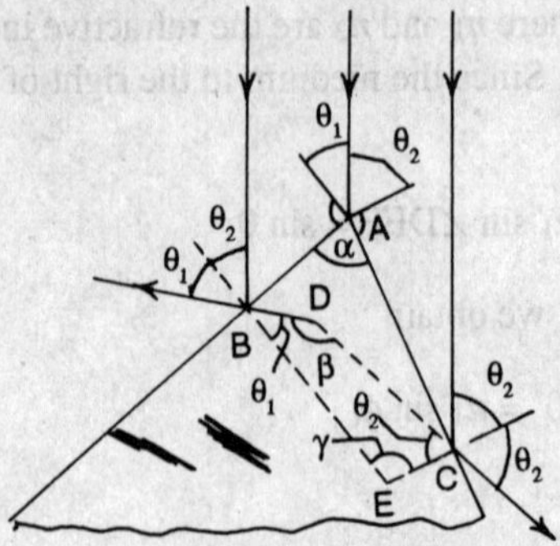

Fig.

The rays striking at B and C are reflected according to the laws of optics, as shown and the angle between the reflected rays is *b*. E is the point at which the normals at B and C meet.

The sum of the angles of quadrilateral ABEC is 360°. Since <ABE and <ACE are each 90°,

$$\alpha + \gamma = 180^\circ \tag{1}$$

In the quadrilateral BDCE,

$$\beta + \gamma + \theta_1 + \theta_2 = 360^\circ \,. \tag{2}$$

Since two of the angles surrounding A are right angles,

$$a + q_1 + q_2 = 180^\circ \tag{3}$$

Add Eqs. (1) and (3)

$$2\alpha + \gamma + \theta_1 + \theta_2 = 360^\circ \tag{4}$$

Subtract (2) from (4)

$$2\alpha - \beta = 0$$

or

$$2\alpha = \beta$$

Thus, the angle between the reflected beams, β, is twice the refracting angle of the prism, α.

Problem 29. A parallel beam of light falls normally on the first face of a prism of small angle α. At the second face it is partly transmitted and partly reflected, the reflected beam striking the first face again and emerging from it in a direction making an angle of 6°30′ with the reversed direction of the incident beam. The refracted beam is found to have undergone a deviation of 1°15′ from the original direction. Calculate the refractive index of the glass and the angle of the prism.

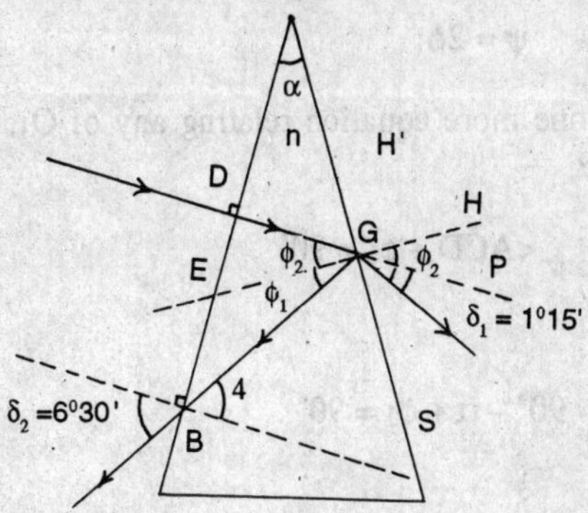

Fig.

Solution: We must first solve for the refractive index, *n*, of the prism glass. Applying Snell's Law to the refraction at point *c* (see figure), we obtain.

$$n \sin \phi_1 = n' \sin \phi_2$$

But n′ is the refractive index of air, which is 1. Hence,

$$n \sin \phi_1 = \sin \phi_2 \qquad (1)$$

Similarly, applying Snell's Law to the refraction at point B

$$n \sin \psi = \sin \delta_2 \qquad (2)$$

Now, we must relate ϕ_1, ϕ_2 and ψ to known quantities.

Note that <HCP = <DCE, since they are vertical angles. Therefore,

$$\measuredangle HCP = \measuredangle DCE$$

$$\phi_2 - \delta_1 = \phi_1$$

or

$$\delta_1 = \phi_2 - \phi_1 \tag{3}$$

Noting that DC and BS are parallel,

$$\psi = 2\phi_1 \tag{4}$$

We need one more equation relating any of O_1, O_2 and ψ to α. $<ACD = 90^\circ - \alpha$.

But $$<ACD + \phi_1 = 90^\circ$$

Hence

$$90^\circ - \alpha + \phi_1 = 90^\circ$$

and

$$\phi_1 = \alpha \tag{5}$$

If all the angles $\left(\phi_1, \phi_2, \psi, \delta_1, \delta_2\right)$ are small, we may approximate the sine of an angle by the angle itself. Using (1) and (2)

$$n\,\phi_1 \approx \phi_2$$

$$n\,\psi \approx \delta_2 \tag{6}$$

Taking the ratio of each equation in (6)

$$\frac{\varphi_1}{\psi} \approx \frac{\varphi_2}{\delta_2}$$

Hence,

$$\varphi_1 \approx \frac{\psi\phi_2}{\delta_2}$$

Using (4)

$$\varphi_1 \approx \frac{2\varphi_1\,\varphi_2}{\delta_2}$$

or

$$\delta_2 = 2\varphi_2 \tag{7}$$

Hence, solving (3) for φ_1

$$\varphi_1 = \varphi_2 - \delta_l \tag{8}$$

Since

$$\varphi = \frac{\delta_2}{2}$$

$$\varphi_1 = \frac{\delta_2}{2} - \delta_1$$

Solving the first equation of (6) for *n*

$$n \approx \frac{\varphi_2}{\varphi_1}$$

with (7) and (8)

$$n = \frac{\delta_2/2}{\frac{\delta_2}{2} - \delta_1} = \frac{1}{1 - \frac{2\delta_1}{\delta_2}}$$

Using the given data

$$n = \frac{1}{1 - \frac{2(1^\circ 15')}{(6^\circ 30')}}$$

$$n = \frac{1}{1 - \frac{2(1\ 1/4^\circ)}{(6\ 1/2^\circ)}}$$

$$n = \frac{1}{1 - \frac{5/2}{13/2}} = \frac{1}{1 - 5/13}$$

$$n = \frac{13}{13 - 5} = \frac{13}{8} = 1.625$$

Furthermore, from (5)

$$\alpha = \varphi_1$$

But, equation (8) tells us that

$$\alpha = \varphi_1 = \frac{\delta_2}{2} - \delta_1 = \frac{6°30'}{2} - 1°15'$$

Hence $\alpha = 3°15' - 1°15' = 2°.$

6

Lenses and Optical Instruments

Problem 1. An object is 24 inches from a convex lens whose focal length is 8 inches. Where will the image be?

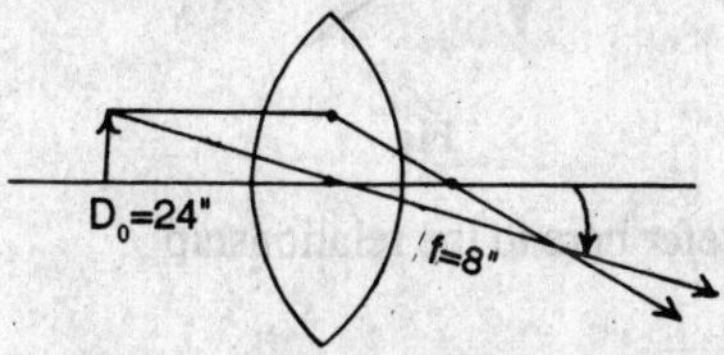

Fig.

Solution: The use of ray diagram is helpful here.

We can see that the image is real (on the far side of the lens and inverted).

We then solve mathematically:

D_O is 24 inches and f is 8 inches. Substituting in

$$\frac{1}{D_O} + \frac{1}{D_I} = \frac{1}{f}$$

$$\frac{1}{24in.} + \frac{1}{D_I} = \frac{1}{8in}$$

Solving

D_I = 12 in.

Hence

$$D_I = 12 \text{ in.}$$

which means that the image is 12 inches from the lens on the side away from the object.

Problem 2. If an object is in a position 8 cm in front of a lens of focal length 16 cm, where and how large is the image? (see figure).

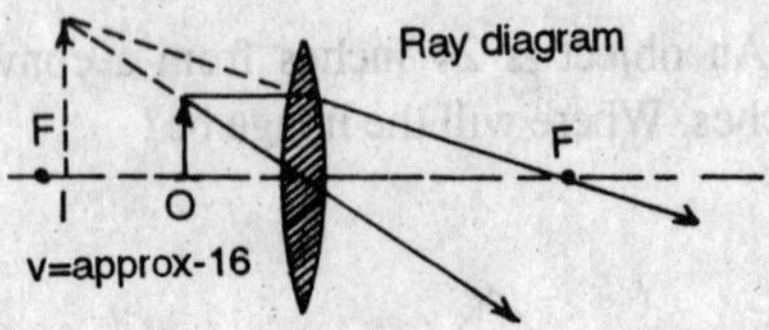

Fig.

Solution: We refer here to the relationship

$$\frac{1}{0} + \frac{1}{i} = \frac{1}{f}$$

Here, the image distance, *i*, is the unknown. The object distance, 0, is 8 cm, while the focal length, *f*, is 16 cm. Thus,

$$\frac{1}{0} + \frac{1}{i} = \frac{1}{f}$$

$$\frac{1}{8cm} + \frac{1}{i} = \frac{1}{16cm}$$

$$\frac{l}{i} = \frac{1}{16cm} - \frac{1}{8cm}$$

$$\frac{l}{i} = -\frac{1}{16cm}$$

$$i = -16\text{cm}.$$

This means that the image is 16 cm in front of the lens.

To calculate the image size, we first calculate the magnification. Thus,

$$m = \frac{-i}{0} = \frac{+16}{8} = 2$$

This means that the image is erect, therefore virtual, and twice as large as the object, or 16 cm high.

Problem 3. An object is 4 inches from a concave lens whose focal length is –12 inches. Where will the image be?

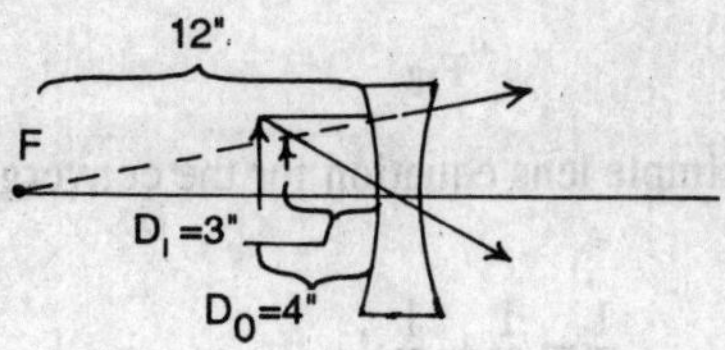

Fig.

Solution: It may be useful to construct a ray-diagram first (see diagram). We draw two rays– one parallel to the axis and one through the centre of the lens.

From the diagram it can be seen that the image is virtual and that it is smaller than the object.

We now attempt a mathematical solution:

Do is 4 inches and f is –12 inches. Substituting in

$$\frac{1}{D_O} + \frac{1}{D_I} = \frac{1}{f}$$

$$\frac{1}{4in.} + \frac{1}{D_I} = \frac{1}{-12in}$$

Solving D_I = – 3 in. Hence D_I = –3 in.

which means that since D_I is negative, the image is 3 inches from the lens on the same side on the lens as the object and is virtual.

Problem 4. A converging lens with a focal length of 3 m forms an image of an object placed 9 m from it. Find the position of the image and the magnification.

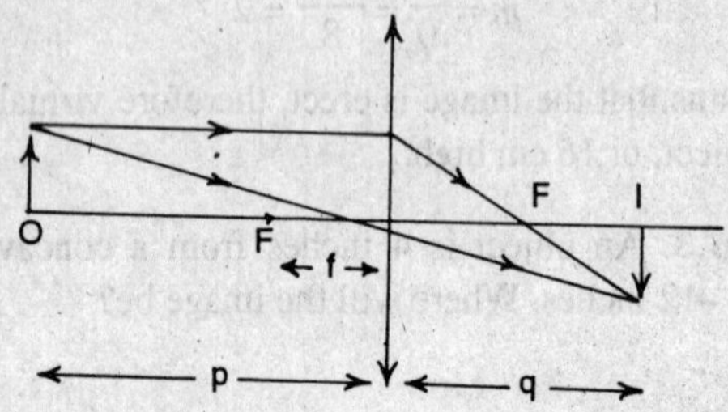

Fig.

Solution: The simple lens equation for the converging lens of this problem is

$$\frac{1}{f} = \frac{1}{p} + \frac{1}{q}$$

where *f*, *p* and *q* are respectively the focal length of the lens and the distance of the object and the image from the lens. The image is real and inverted (see the figure). Substituting the given values in the above equation, we get

$$\frac{1}{q} = \frac{1}{f} - \frac{1}{p} = \frac{1}{3m} - \frac{1}{9m} = \frac{2}{9}m^{-1} \quad \text{and}$$

$$q = \frac{9}{2}\,\text{m} = 45\text{ m}.$$

Since the value of q is positive, the image occurs on the right side of the lens. The magnification M is

$$M = \frac{q}{p} = \frac{45m}{9m} = 0.5$$

so the image is one-half as high as the object.

Problem 5. A converging lens of 5.0 cm focal length is used as a simple magnifier, producing a virtual image 25cm from the eye. How far from the lens should the object be placed? What is the magnification?

Solution: The image produced is virtual, on the same side of the lens as the object. Therefore, the distance q of the image from the lens is negative. Using the general lens equation

$$\frac{1}{p}+\frac{1}{q}=\frac{1}{f}; \quad \frac{1}{p}=\frac{1}{f}-\frac{1}{q}.$$

Solve for p and substitute values.

$$p=\frac{fq}{q-f}=\frac{5.0\,cm\,(-25\,cm)}{-25cm-5.0cm}=4.2\text{ cm.}$$

Using absolute values to find the magnification,

$$M=\left|\frac{q}{p}\right|=\left|\frac{-25\,cm}{4.2cm}\right|=5.9.$$

Since the image is on the same side of the lens as the object, the light has been focused in a way that produces an erect image. In a real image, which is seen on the opposite side of the lens in relation to the object, the object is inverted.

Problem 6. The length of a microscope tube is 10 centimetres. The focal length of the objective is 0.5 centimetres and the focal length of the eyepiece is 2 centimetres. What is the magnifying power of the microscope?

Solution: The magnification of this compound microscope is equal to the product of the magnifications of its component lenses or

$$M = M_1 \times M_2$$

which in this case is

$$\frac{25cm}{f_O}\times\frac{10cm}{f_E}.$$

We use 25 cm since in most cases we want the image to be formed at that distance from the ocular. Therefore, the solution is:

$$\frac{25L}{f_O f_E}=\frac{25\times 10}{0.5\times 2}=250$$

Problem 7. The lens system of a certain portrait camera may be taken as equivalent to a thin converging lens of focal length 10.0 in. How far behind the lens should the film be located to receive the image of a person seated 50.0 in. from the lens? How large will the image be in comparison with the object?

Solution: The general equation for thin lenses states

$$\frac{1}{p}+\frac{1}{q}=\frac{1}{f}$$

where p is the distance of the object, in this case the person, in front of the lens, and q is the distance between the image formed and the lens. The focal length is represented by f. All distances are measured with respect to the optical centre of the lens. This centre is defined as the point through which light can pass without being bent. Since the lens is a converging one, it is convex. This means that it is thicker at the centre than at the edges. Substitute values in the above equation noting that q is positive when it is on the opposite side of the lens in relation to the object.

$$\frac{1}{q}=\frac{1}{10}-\frac{1}{50}=\frac{50-10}{(10)(50)}=\frac{4}{50}$$

$$q=\frac{50}{4}=12.5 \text{ in.}$$

To find the magnification M, the equation $M=\frac{q}{p}$ can be used.

Substituting,

$$M=\frac{12.5\ in}{50.0\ in.}=0.250.$$

The image will be one-fourth as large as the object.

Problem 8. In one of Edgar Allan Poe's stories, the author

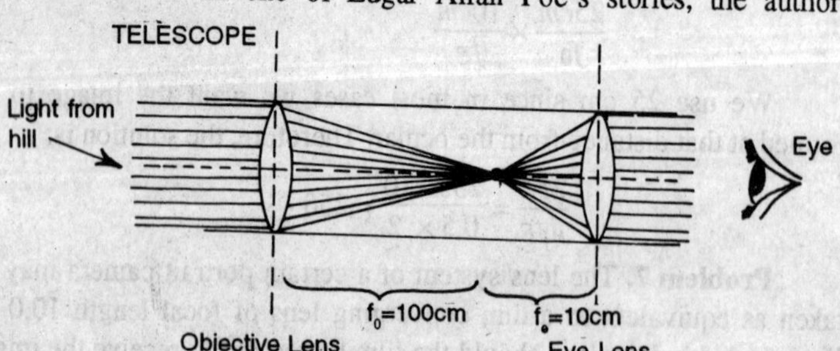

Fig.

describes his terrifying experience of focusing a telescope on a distance hill and observing a "dragon" crawling up it. The punch line comes when he realizes that the dragon is an ant crawling up the windowpane through which he is observing the hill. Explain why this is impossible.

Solution: The distant hill is, for all practical purposes, at infinity. The objective lens of the telescope forms an image of it at its second focal point, which is also the first focal point of the eye lens (see figure). The final image is being viewed at infinity. Unless the ant is at infinity or at the second focal point of the objective lens, i.e., unless it occupies the same position as the object or the intermediate image, its final image cannot be at infinity also, and thus it will not be viewed at the same position as the image of the hill.

In fact, assuming that the telescope has focal lengths of 100cm and 10 cm for its objective and eye lenses, respectively, and that the ant is 200 cm from the objective, we can work out where the final image of the ant lies.

Treating the objective, we find that

$$\frac{1}{s_1}+\frac{1}{s_i}=\frac{1}{f_1}$$

where s_i and s_i are the object and image distances, respectively.

$$\frac{1}{200cm}+\frac{1}{s_i}=\frac{1}{100cm}$$

$$\frac{1}{s_i{}'}=\frac{1}{200cm}-\frac{1}{200cm}=\frac{1}{200cm}$$

$$s_i{}'=200\text{ cm.}$$

The objective lens and eye lens are 110 cm apart (see figure) and the image formed by the objective will act as a virtual object for the eye lens. The object distance is $s_2 = (110 - 200)$ cm $= -90$ cm.

$$-\frac{1}{90cm}+\frac{1}{s_2{}'}=\frac{1}{10cm}$$

$$\frac{1}{s_2{}'}=\frac{9}{90cm}+\frac{1}{90cm}=\frac{10}{90cm}$$

or $s_2' = 9$cm.

The final image is real one on the observer's side of the eye lens and therefore Poe, when he looked through the telescope, could not see an image of the ant.

Problem 9. A certain farsighted person has a minimum distance of distinct vision of 150 cm. He wishes to read type at a distance of 25 cm. What focal-length glasses should he use?

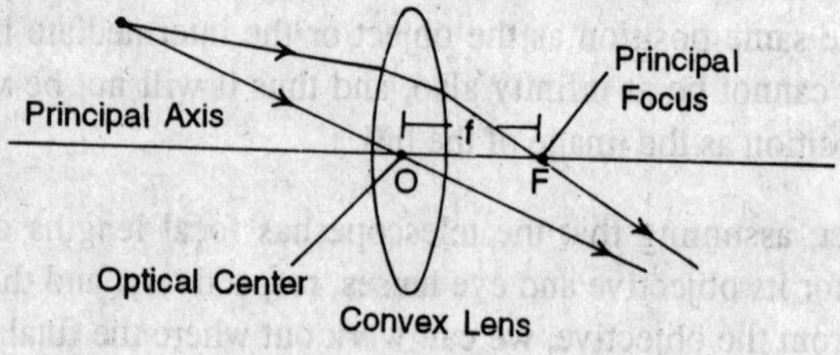

Fig.

Solution: The principal axis is the line passing through the centres of curvature of the faces of the lens. The optical centre is the point in the lens through which light can pass without being bent. All rays of light parallel to the principal axis pass through F, the point of principal focus. The distance *f* between F and the optical centre is called the focal length of the lens. It is positive for converging, convex lenses, negative for concave, diverging lenses.

Since the person cannot see clearly objects closer than 150 cm, the lens must form a virtual image at that distance. Since the image is formed on the same side of the lens as the object, its distance *q* from the lens is negative.

$$p = 245 \text{ cm}$$

$$q = -150 \text{ cm.}$$

Substituting in the general equation for lenses,

$$\frac{1}{p} + \frac{1}{q} = \frac{1}{f}$$

$$\frac{1}{25cm} + \frac{1}{-150cm} = \frac{1}{f}$$

$$f = 30 \text{ cm.}$$

Since f is positive, this lens is converging and is convex.

Problem 10. A microscope has an objective lens of 10.0 mm focal length and an eyepiece of 25.0 mm focal length. What is the distance between the lenses, and what is the magnification if the object is in sharp focus when it is 10.5 mm from the objective?

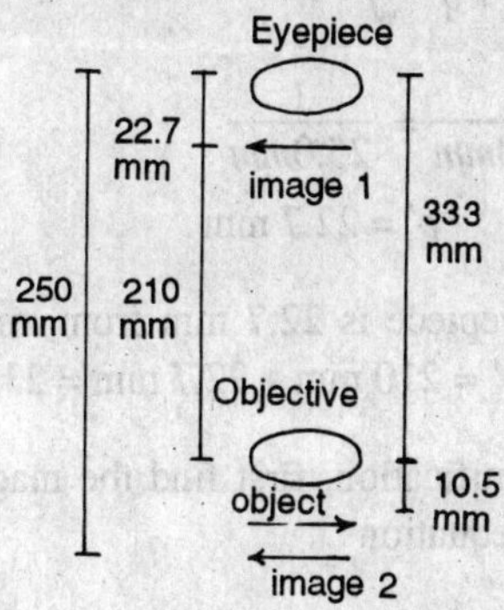

Fig.

Solution: This is a compound microscope which uses converging lenses to produce large magnification. A short-focus lens, the objective, is placed near the object. It produces a real, inverted and magnified image. The eyepiece further magnifies this inverted image and produces a virtual image 250 mm from the eyepiece. This is the distance of most distinct vision.

Considering the image produced by the objective, we use the optics equation

$$\frac{1}{p} + \frac{1}{q} = \frac{1}{f}$$

p represents the distance of the object from the objective lens, q the distance of image 1 from the lens, and f is the focal length. Substituting.

$$\frac{1}{10.5cm} + \frac{1}{q} = \frac{1}{10.0mm}$$

$$q = 210 \text{ mm}$$

Therefore image 1 is 210 mm from the objective lens, and is real.

The eyepiece magnifies image 1 and produces a second virtual image 250 mm from the lens so as to provide most distinct vision. Since image 2 is virtual, on the same side of the lens as image 1, the distance q' of image 2 from the eyepiece is negative. Using the optics equation again.,

$$\frac{1}{p'} + \frac{1}{q} = \frac{1}{f}$$

$$\frac{1}{p'} + \frac{1}{-250mm} = \frac{1}{25.0mm}$$

$$p' = 22.7 \text{ mm}$$

Therefore the eyepiece is 22.7 mm from image 1. The distance between lenses is $q + p'$ = 210 mm + 22.7 mm = 233 mm = 23.3 cm

To find the magnification, first find the magnification produced by each lens using the equation

$$M = \frac{q}{p}$$

Magnification by objective:

$$M_O = \frac{210mm}{10.5mm} = 20.0$$

Magnification by eyepiece:

$$M_e = \frac{250mm}{22.7mm} = -11.0$$

Total magnification:

$$M = M_e M_O = -11.0 \times 20.0 = -220$$

Problem 11. A projector with a lens of 40 cm focal length illuminates an object slide 4.0 cm long and throws an image upon a screen 20 m from the lens. What is the size of the projected image?

Solution: We use the lens formula $\frac{1}{p} = \frac{1}{f} - \frac{1}{q}$ where p is the object distance, f is the focal length of the lens, and q is the image distance. Upon substitution of the given values in the formula, we have

$$\frac{1}{p} = \frac{1}{40cm} - \frac{1}{20m}$$

In order to do the substraction, we must change 20 m to centimetres. Since 1m = 10^2 cm, 20 cm = 2000 cm and

$$\frac{1}{p} = \frac{1}{40cm} - \frac{1}{2000cm}$$

$$\frac{1}{p} = \frac{2000cm - 40cm}{8000cm^2} = \frac{1960}{80000cm}$$

$$p = \frac{80000cm}{1960} = 40.8 \text{ cm}$$

This is the object distance.

The magnification M of the lens is defined as

$$M = \frac{q}{p} = \frac{2000cm}{40.8cm} = 49.$$

Given, then, an object 4 cm long, the size of its projected image is

Image length = M × object length

$$= 49 \times 4 \text{ cm} \approx 200 \text{ cm}.$$

Problem 12. The frames in a home movie must be magnified 143 times before the picture formed on a screen 12 ft from the projection lens is large enough to please the family watching. What distance must the film be from the lens and what is the focal length of the lens?

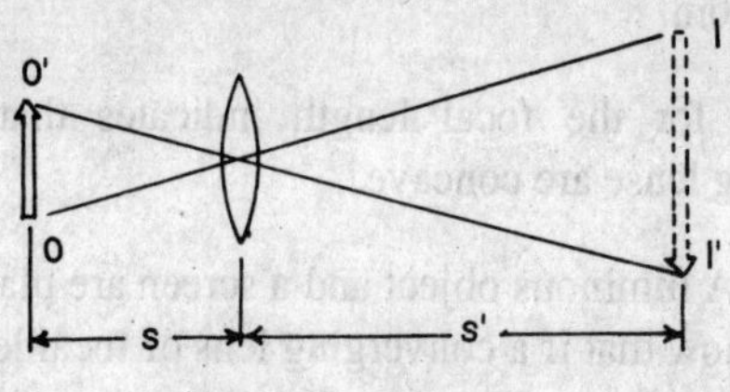

Fig.

Solution: The magnification produced by the lens is given by

$$m = -\frac{s'}{s} = -\frac{12ft}{s} = -143.$$

where s' and s are the image and object distance, respectively. (The image must be inverted that is, the magnification is negative, since s must be positive (see figure). Thus the film-to-lens distance is

$$s = \frac{12}{143}\text{ ft} = \frac{144}{143}\text{ in.} = 1.007\text{ in.}$$

Applying the lens formula, we can obtain the focal length, since

$$\frac{1}{f} = \frac{1}{s} + \frac{1}{s'} = \frac{143}{12ft} + \frac{1}{12ft} = \frac{144}{12ft}$$

$$f = \frac{12}{144} ft = 1\text{ in.}$$

Problem 13. When an object is placed 20 in. from a certain lens, its virtual image is formed 10 in. from the lens. Determine the focal length and character of the lens.

Solution: Since the image is virtual, on the same side of the lens as the object, its distance from the lens, q, is negative. Substitution in the general equation for lenses yields

$$\frac{1}{p} + \frac{1}{q} = \frac{1}{f}$$

$$\frac{1}{20in} + \frac{1}{-10in} = \frac{1}{f}$$

$$\frac{1}{f} = \frac{-10\ in\ +\ 20\ in}{(20\ in) \times (-10\ in)} = -\frac{10}{200in}$$

$$f = -20\text{in.}$$

The negative sign for the focal length indicates that the lens is diverging. Diverging lense are concave.

Problem 14. A luminous object and a screen are placed at a fixed distance D apart. Show that if a converging lens of focal length f, where $f <$ D/4, is inserted between them it will produce a real image of the object on the screen for two positions separated by a distance $d = \sqrt{D - (D - 4f)}$, that the ratio of the two imge sizes for these two positions of the lens is $(D - d)^2 / (D + d)^2$.

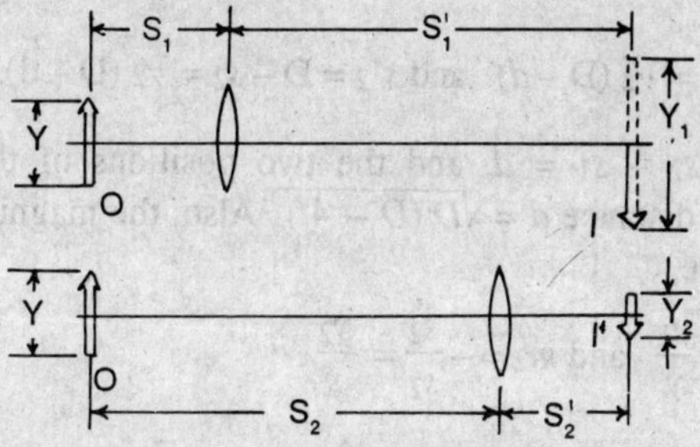

Fig.

Solution: In both cases,

$$\frac{1}{s}+\frac{1}{s}=\frac{1}{f}$$

where s is the object distance, and s′ is the image distance.

$$\frac{ss'}{s+s'}=f \quad \text{or} \quad \frac{ss'}{D}=f.$$

since s + s' = D (see figure)

$$ss' = ft \qquad (1)$$

and $s + s' = D$ (2)

Let d = $s - s'$

$$d^2 = (s - s')^2 = (s + s')^2 - 4s' = D^2 - 4fD.$$

Therefore $d = \sqrt{D\,(D - 4f)}$. Also, from Eqs. (1) and (2),

$$s + \frac{fD}{s} = D \qquad \text{or} \qquad s^2 - Ds + fD = 0.$$

$$s = D \pm \frac{\sqrt{D^2 - 4fD}}{2} = \frac{D \pm d}{2}$$

This is a general formula for s. It has 2 roots, one of which must be s_1, and the other of which must be s_2. (see figure).

$$s_1 = ½\,(D - d) \text{ and } s_2 = ½\,(D + d)$$

From (2)

$s'_1 = D - s_1 = ½ (D - d)$ and $s'_2 = D - s_2 = ½ (D - d)$.

Therefore $s_2 - s_1 = d$, and the two positions of the lens are separated by the distance $d = \sqrt{D(D-4f)}$. Also, the magnification for the 2 positions are

$$m_1 = \frac{s'_1}{s_1} = \frac{y_1}{y} \text{ and } m_2 = -\frac{s'_2}{s_2} = \frac{y_2}{y}$$

$$\therefore \frac{y_2}{y_1} = \frac{m_2}{m_1} = \frac{s_1 s'_2}{s'_1 s_2} = \frac{¼ (D-d)^2}{¼ (D+d)^2} = \frac{(D-d)^2}{(D+d)^2}.$$

Note that this method only works if $f < D/4$; for otherwise d^2 is negative and d is thus imaginary.

PHOTOMETRY

Problem 15. What is the illumination 5 feet from a 75 candlepower lamp?

Solution: Illumination is defined as flux divided by area $\left(E = \frac{F}{A}\right)$.

Making the simplifying assumption that our lamp is a point source, the total flux that it emits is 4πI where I is the intensity of the source. If we construct a sphere enveloping the point source with the point source at its centre, and having a radius of 5 ft., we can find the illumination at this distance.

Hence,

$$E = \frac{F}{A} = \frac{4\pi I}{4\pi r^2} = \frac{I}{r^2}$$

$$\text{And Illumination} = \frac{75 \text{ candlepower}}{(5 \times 5) \text{ ft}^2}$$

$$= 3 \text{ foot-candles.}$$

Problem 16. A spotlight equipped with a 32-candle bulb concentrates the beam on a vertical area of 125 ft^2 at a distance of 100 ft. What is the luminous intensity of the spot of light in the beam direction?

Solution: In a lamp, electrical power is supplied and radiation is emitted. The radiant energy emitted per unit of time is called the radiant flux. Only a fraction of this lies within the wavelength interval (400 mμ to 700 mμ) which can produce a visual sensation in the human eye. The part of the radiant flux which affects the eye is called the luminous flux F and is measured in lumens. The luminous intensity I of a source is defined as the flux it emits per unit solid angle ($I = F/\omega$). The solid angle ω is a measure of the size of the cone of radiation and is defined in steradians as

$$\omega = \frac{A}{R^2}$$

where R is any distance and A is the area of a spherical surface surrounding the source which is irradiated by the light source at this radius R.

Assuming that the spotlight radiates in all directions, the total area illuminated at a distance R is equal to the area of a sphere of radius R or $4\pi R^2$. Therefore the total solid angle

$$\omega_{total} = \frac{4\pi R^2}{R^2} = 4\pi \text{ sterad.}$$

and the total flux emitted by the bulb is given by

$$F_{total} = I\,\omega_{total} = (32 \text{ candles})\,(4\pi \text{ steradians}) = 400 \text{ lu}$$

This total flux is concentrated into a beam with a small solid angle through the use of a reflector and lens. This solid angle is

$$\omega = \frac{A}{R^2} = \frac{125 ft^2}{(100\,ft)^2} = 0.0125 \text{ sterad.}$$

The intensity of this beam is then

$$I = \frac{F_{total}}{\omega} = \frac{400\ lu}{0.0125\ sterad} = 32{,}000 \text{ candles.}$$

Problem 17. A "point-source" unshaded electric lamp of luminous intensity 100 candles is 4.0 ft above the top of a table. Find the illuminance of the table (a) at a point directly below the lamp and (b) at a point 3.0 ft from the point directly below the lamp.

Solution: (a) Illuminance E is the luminous flux F incident per unit area A. Then

$$E = \frac{F}{A}$$

Also $\qquad F = I\omega$

and $\qquad \omega = \frac{A^{\circ}}{R^2}$

If the area A is perpendicular to the path of radiation from the point source, the illuminance can also be expressed as

$$E = \frac{F}{A} = \frac{I\omega}{A} = \frac{(A/R^2)}{A} = \frac{I}{R^2}$$

which is the equation needed since the given quantities are the luminous intensity I and the distance R from the point source. Furthermore, we know that the area is perpendicular to the radiation being emitted by the lamp.

$$E = \frac{I}{R^2} = \frac{100 \text{ candles}}{(4.0 \text{ ft})^2} = 6.25 \text{ lu/ft}^2$$

(b) For this second point, the area illuminated is a distance

$$R = \sqrt{(4.0ft)^2 + (3.0ft)^2} = 5.0 \text{ ft}$$

from the lamp, which was found by using the trigonometric relation for a right triangle. The illuminance is

$$E = \frac{F}{A}$$

where $\qquad F = I\omega.$

The solid angle ω is defined as the ratio of the area upon which the source radiates at a radius R to this distance R. The area upon which the source radiates for this can be seen from the figure to be A cos θ. Therefore,

$$\omega = \frac{A \cos \theta}{R^2}$$

and $$E = \frac{F}{A} = \frac{I\omega}{A} = \frac{I\,(A\cos\theta/R^2)}{A} = \frac{I\cos\theta}{R^2}$$

Using trigonometry, cos θ is found to be

$$\cos\theta = \frac{4.0\,ft}{5.0\,ft} = 0.80$$

Then the illuminance on the area A is

$$E = \frac{I\cos\theta}{R^2} = \frac{(100\ candles)\ (0.80)}{(5.0\,ft)^2} = 3.2\ \text{lu/ft}^2$$

Problem 18. A standard 48-candle lamp is placed 36 in. from the screen of a photometer and produces there the same illuminance as a lamp of unknown intensity placed 45 in. away. What is the luminous intensity of the unknown lamp?

Solution: Illuminance E is equal to the ratio of the luminous intensity I to the square of the distance R from the source. Since the illuminance of the two lamp is equal,

$$E_1 = E_2$$

where the subscripts 1 and 2 refer to the lamps at distances of 45 in. and 36 in. respectively. Then

$$\frac{I_1}{R_1} = \frac{I_2}{R_2}$$

and substituting values,

$$\frac{I_1}{(45\ in)^2} = \frac{48\ candles}{(36\ in)^2}$$

$$I_1 = \frac{(45in)^2\ (48\ candles)}{(36in)^2} = 75\ \text{candles}$$

Note that the distance can be expressed in any unit as long as they are consistent.

7

Optical Interference

Problem 1. Derive relation describing the interference effects observed when light is reflected from a thin film. (see figure)

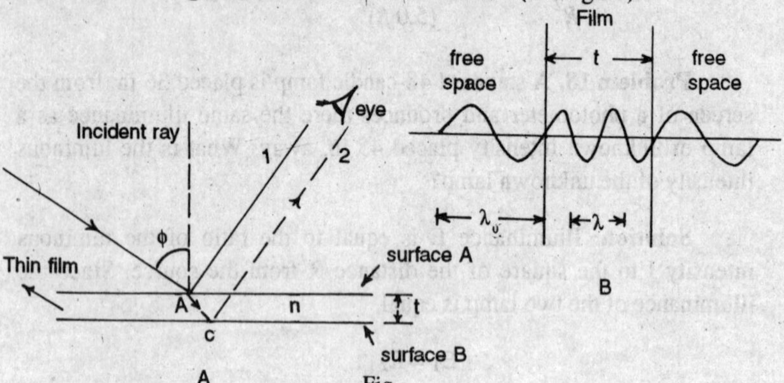

Fig.

Solution: In order to understand the interference effects produced by a thin film, we trace the path of an incident ray of light, as shown in figure (a). The incident ray first encounters surface A, where it is partially reflected and partially absorbed. Since the refractive index of the film is greater than the refractive index of air, the reflected ray undergoes a 180° phase change. The transmitted part of the incident ray now encounters interface B, where it is partially reflected and partially absorbed. However, this time the reflected ray undergoes no phase change since it is travelling from a region of high refractive index to a region of low refractive index. Hence, rays 1 and 2 differ in phase by 180°, or $\lambda_0/2$ where λ_0 is the wavelength of the incident light in free space.

In addition to the 180° phase change due to reflection, ray 2 travels a distance $2t$ greater than ray 1. (This holds only if the rays shown are incident at angle f which is very small). Then, if we want to observe destructive interference, $N\lambda$, where N = 0, 1, 2, ... But, λ is not the wavelength of the light in free space, but rather, the wavelength of light in the film (see figure (b)). However, the wavelengths λ_0 and λ are related. By definition of the refractive index of the film.

$$n = c/v$$

where c is the speed of light in free space, and v is its speed in the film. If λ_0 and f_0 are the free space wavelength and frequency of light, we may write.

$$c = \lambda_0 f_0$$

Similarly,

$$v = \lambda f$$

where λ and f are the wavelength and frequency of light inthe film. But the frequency of light is the same in all media. Then

$$v = \lambda f_0$$

Hence

$$n = \frac{\lambda_0 f_0}{\lambda f_0} = \frac{\lambda_0}{\lambda}$$

and

$$\lambda_0 = n\lambda$$

Combining this fact with the previous discussion, we obtain

$$2t = N\lambda = \frac{N\lambda_0}{n} \qquad N = 0, 1, 2, \ldots \qquad \text{destructive interference}$$

$$2t = (N + ½)\lambda = \frac{(N + ½)\lambda_0}{n} \qquad N = 0, 1, 2, \ldots \qquad \text{constructive interference}$$

or

$$t = \frac{N\lambda_0}{2n} \qquad N = 0, 1, 2, \ldots \qquad \text{destructive interference}$$

$$t = \frac{(N+½)\lambda_0}{2n} \qquad N = 0, 1, 2, \ldots \qquad \text{constructive interference}$$

Problem 2. In a double slit interference experiment the distance between the slits is 0.05 cm and the screen is 2 meters from the slits. The light is yellow light from a sodium lamp and it has a wavelength of 5.89×10^{-5} cm. What is the distance between the fringes?

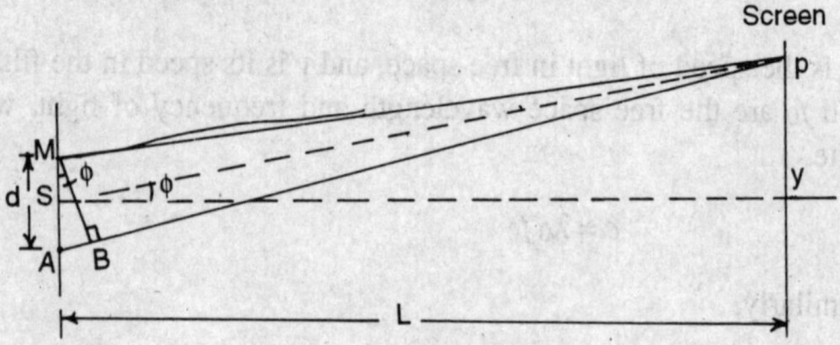

Fig.

Solution: To find the distance between fringes in the double slit experiment, we must first derive the formulas for the location of the maxima and minima of the fringe pattern. Let us examine this experiment in more detail.

Light is incident on the 2 slits from the left. (See figure) MP and AP represents 2 rays of light, one from each slit, arriving at P. Typically, L >> d, and we may consider MP to be equal to BP. Assuming that the light rays emerging from the slits are in phase, the two light rays arriving at P will be out of phase because light from A must travel the extra distance AB when compared with light from M. If this path difference (AB = $d \sin \phi$) is equal to an even number of half wavelengths, P will be a maximum point. If AB equals an odd number of half wavelengths, P will be a minimum point. Hence,

For a maximum $\sin \phi = (2n)\dfrac{\lambda}{2d}$ (n = 0, 1, 2, . . .)

For a minimum $\sin \phi = (2n+1)\dfrac{\lambda}{2d}$ (n = 0, 1, 2, . . .)

Therefore, the angular location of adjacent maxima on the screen (say the nth and n+1) *th* maxima,) is

$$\sin(\phi_{n+1}) = \frac{(2(n+1)\lambda}{2d} = \frac{(n+1))\lambda}{d}$$

$$\sin(\phi_n) = \frac{(2n)\lambda}{2d} = \frac{n\lambda}{d}$$

But, if ϕ is small,

$$\sin(\phi_{n+1}) \approx \tan(\phi_{n+1})$$

$$\sin(\phi_n) \approx \tan(\phi_n).$$

Hence, using (1) and the figure,

$$\frac{Y_{n+1}}{L} = \frac{(n+1)\lambda}{d}$$

$$\frac{Y_n}{L} = \frac{n\lambda}{d}$$

hence,

$$Y_{n+1} - Y_n = \frac{(n+1)\lambda L}{d} - \frac{n\lambda L}{d}$$

$$Y_{n+1} - Y_n = \frac{\lambda L}{d}.$$

This is the screen separation of 2 adjacent maxima.

If $\lambda = 5.89 \times 10^{-5}$ cm

$L = 200$ cm

$d = 0.05$ cm.

$$Y_{n+1} - Y_n = \frac{(5.89 \times 10^{-5} \times 200)cm^2}{0.05cm}$$

$$= .233 \text{ cm.}$$

Problem 3. With two slits spaced 0.2 on m apart, and a screen at a distance of $l = 1$m, the third bright fringe is found to be displaced h = 7.5 mm from the central fringe. Find the wavelength λ of the light used. See the figure on next page.

Solution: When the difference D between the path lengths of the

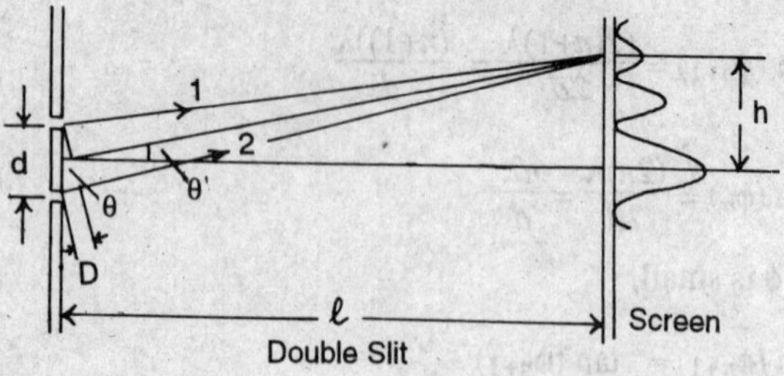

Fig.

rays 1 and 2 is an integral multiple of the wavelength λ, one obtains a maximum (bright fringe) of the interference pattern on the screen. From the figure we see that

$$D = d \sin \theta .$$

If l is much larger than the distance between the two slits, we see that $\theta' = \theta$, where θ' relates the position of the maximum on the screen to the distance between slits and the screen,

$$\tan \theta' = \frac{h}{l} .$$

The approximation $l >> d$ also means that q is small; for which case we have

$$\tan \theta' = \sin \theta' .$$

Therefore, for the third maximum to occur at h, D must be 3λ;

$$D = d \sin \theta = 3\lambda$$

or

$$\sin \theta = \sin \theta' = \frac{3\lambda}{d}$$

$$\frac{h}{l} = \frac{3\lambda}{d}$$

which gives λ as

$$\lambda = \frac{dh}{3l}$$

$$= \frac{0.75\ cm \times 0.02 cm}{3 \times 100 cm} = 5 \times 10^{-5}\text{ cm}$$

$$= 500 \times 10^{-9}\text{ m} = 500\text{ mm.}$$

Problem 4. In a double-slit experiment, D = 0.1 mm and L = 1m. If yellow light is used, what will be the spacing between adjacent bright lines?

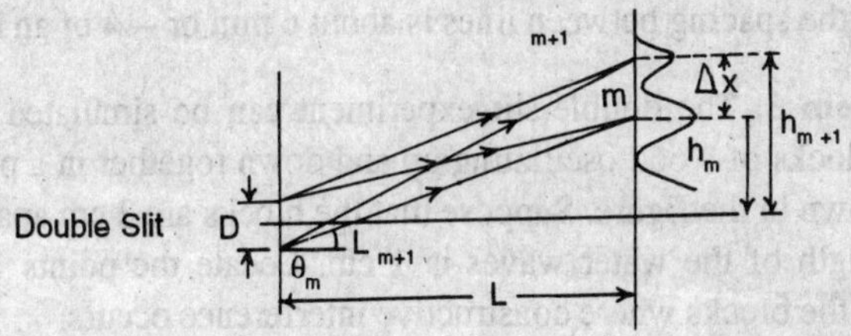

Fig.

Solution: The wavelength of yellow light is approximately 6 × 10^{-5} cm. Let the separation between the (*n*+1)th and m′ th maxima be Δx, as shown in the figure, Then,

$$\sin \theta_m = \frac{m}{D}\lambda$$
$$\sin \theta_{m+1} = \frac{m+1}{D}\lambda$$

The angles θ_m and θ_{m+1} are related to the positions h_m, and h_{m+1} of these maxima on the screen;

$$\sin \theta_m = \frac{h_{m+1}}{L}$$

$$\sin \theta_{m+1} = \frac{h_m}{L}$$

hence,

$$\sin \theta_{m+1} - \sin \theta_m = \frac{\lambda}{D}$$

$$\frac{h_{m+1} - h_m}{L} = \frac{\lambda}{D}$$

$$\frac{\Delta x}{L} \approx \frac{\lambda}{D}$$
$$\Delta x = \frac{L}{D}$$

$$= \frac{(6 \times 10^{-5} cm) \times (100 cm)}{10^{-2}\, cm}$$

$$= 0.6 \text{ cm}$$

Thus, the spacing between lines is about 6 mm or –¼ of an inch.

Problem 5. The double-slit experiment can be simulated using two small blocks of wood oscillating up and down together in a pool of water as shown in the figure. Suppose that the blocks are 3 cm apart and the wavelength of the water waves is 1 cm. Locate the points 10 cm from one of the blocks where constructive interference occurs.

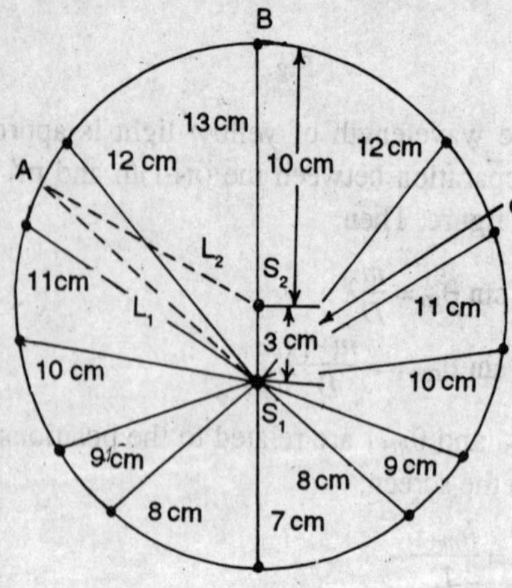

Fig.

Solution: The motion of the blocks produce circular waves which are emanating from two sources S_1 and S_2 separated by d = 3 cm. In order to observe the interference pattern that forms at 10 cm from one of the blocks, draw a circle of radius 10 cm around source S_2 as shown in the figure.

Constructive interference occurs at points where the difference in the distances to the two blocks in either zero or some whole number of wavelengths.

Let A be a point on the circle at a distance L_1 from S_1. All the points on the circle are L_2 = 10 cm from S_2. Constructive interference occurs at points for which the length $|L_1 - L_2|$ is an integer multiple of the wavelength $\lambda = 1$ cm. Therefore, we have constructive interference at points on the circle that are

$$L_1 = 10\text{ cm} - 3\text{cm} = 7\text{ cm}$$

$$= 10\text{ cm} - 2\text{ cm} = 8\text{ cm}$$

$$= 10\text{ cm} - 1\text{ cm} = 9\text{ cm}$$

$$= 10\text{ cm}$$

$$= 10\text{ cm} + 1\text{ cm} = 11\text{ cm}$$

$$= 10\text{ cm} + 2\text{ cm} = 12\text{ cm}$$

$$= 10\text{ cm} + 3\text{ cm} = 13\text{ cm}$$

from S_1.

Problem 6. A student sets up a Young's experiment using a sodium vapor lamp as a light source and placing his slits 1 m from a screen. He is unsure of the slit separation and, standing beside the slits, he varies the separation and finds the interference pattern vanishes if the slits are too far apart. If the angular resolution of his eye is 1 minute of arc, how far apart are the slits when he just cannot see that interference fringes?

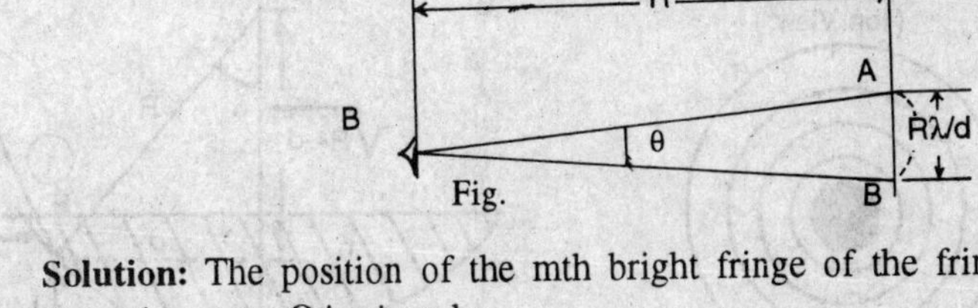

Fig.

Solution: The position of the mth bright fringe of the fringe system from the centre O is given by

$$y_m = \frac{mR\lambda}{d},$$

where λ is the wavelength used (see figure (A)). Similarly, for the (m + 1) th fringe,

$$y_{m+1} = \frac{(m+1)R\lambda}{d}$$

Thus, the fringe separation is

$$y_{m+1} - y_m = \frac{R\lambda}{d}$$

a quantity independent of *m* and thus constant throughout the fringe system.

The angle subtended by this fringe separation at the student's eye is (see figure (B))

$$\theta = \frac{R\lambda/d}{R} = \frac{\lambda}{d}\text{ rad.}$$

Here, we have approximated the arc shown in figure (B) (dotted line) by the straight line distance $\overline{AB}$.

But the student cannot see the fringes as distinct unless the angle neighbouring fringes subtend at his eye ≥ 1 minute of arc. But 1 minute of arc is 1/60° = π/60 × 180 rad. Hence,

$$\theta_{min} = \frac{\lambda}{d_{max}} = \frac{\pi}{180 \times 60}\text{ rad}$$

$$\therefore\ d_{max} = \frac{180 \times 60 \times 5.89 \times 10^{-5}}{\pi\ rad} = 2.025\text{ mm.}$$

Problem 7. When a flat plate of glass and a lens are placed in

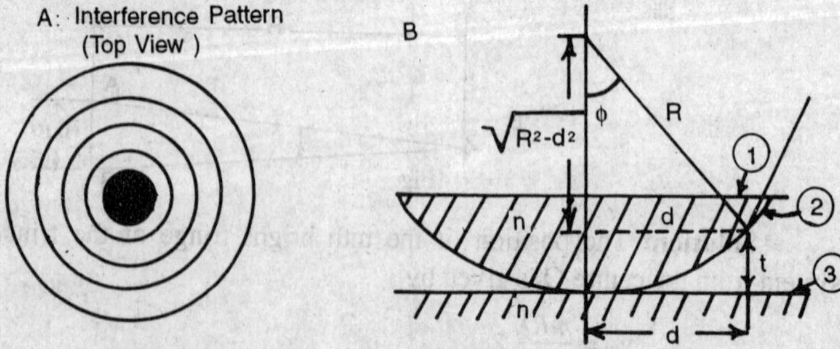

Fig.

contact, a distinctive interference pattern, known as Newton's Rings, is observed. (See figure A). Derive a formula giving the location of the fringes of the interference pattern relative to the centre of the lens. (See figure B).

Solution: Destructive interference will result when the waves reflected from the apparatus shown in figure (B) are 180° out of phase. Let us trace the path of an incident ray. The ray will be partially reflected and partially transmitted at surface (1). When a ray of light is tramsmitted from a region of low refractive index to a medium of higher refractive index, it undergoes a phase change of 180°. Hence, the ray reflected at surface (1) is 180° out of phase with the incident ray. The transmitted ray next encounters surface (2). At this surface there is no phase change, since the light leaves an area of high refractive index and enters a region of low refractive index. In addition, part of this light is reflected at surface (2). The light transmitted at surface (2) next encounters surface (3) and is reflected with a 180° or $\lambda/2$ phase change. (λ is the wavelength the light). Hence, the ray reflected at (2), and the ray reflected at (3) are 180° out of phase.

Now, the ray reflected at (3) travels a distance $2t$ greater than the ray reflected at (2). We will see destructive interference whenever $2t$ is an integral number of wavelengths, since the additional $\lambda/2$ required for destruction is provided by the phase change due to reflection. Hence

$2t = n\lambda \qquad n = 0, 1, 2, \ldots$ destructive interference (1)

$2t = (n + ½)\lambda \quad n = 0, 1, 2, \ldots$ constructive interference

We must now find the location of the interference fringes in terms of the geometry of figure (B).

$$t = \mathrm{R} - \sqrt{R^2 - d^2} \tag{2}$$

$$t = \mathrm{R} - \mathrm{R}\sqrt{1 - d^2/R^2}$$

$$t = \mathrm{R}\left\{1 - \sqrt{1 - d^2/R^2}\right\} \tag{3}$$

But $d << \mathrm{R}$ and $d/\mathrm{R} << 1$. (This means that the radius of curvature of the lens is large). We may therefore approximate the square root in

(3) by the binomial theorem. Therefore.,

$$\sqrt{1 - d^2/R^2} \approx 1 - d^2/2\,R^2 \qquad (4)$$

Substituting (4) in (3)

$$t = R\,(1 - 1 + d^2/2\,R^2)$$

$$t = d^2/2R$$

$$d = \sqrt{2tR}$$

Using (1)

$$d = \sqrt{n\lambda R} \qquad n = 0, 1, 2, \ldots$$

destructive interference

$$d = \sqrt{(n + ½)\lambda R} \quad n = 0, 1, 2, \ \ldots$$

constructive interference

The first equation locates the dark rings relative to the centre of the lens, and the second equation locates the bright rings.

Problem 8. An interferometer illuminated with red light from cadmium (λ = 6438 A) is used to measure the distance between two points. Calculate this distance, D, if 120 minima pass the reference mark as the mirror is moved from one of the points to the other.

Solution: The fringes observed represent the interference of light rays. Consider 2 rays of light, initially in phase. (That is, they interfere to produce areas of high light intensity, or maxima). If the phase of one ray is varied until a minima is observed, we will find that the phase difference of the 2 rays is now $\frac{\lambda}{2}$, where λ is the wavelength of the light. Hence, 2 minima are separated by l. Therefore,

$$D = N\,\lambda = 120\,(6.438 \times 10^{-5}\text{ cm}) = 0.00773\text{ cm.}$$

Problem 9. To produce a minimum reflection of wavelengths near the middle of the visible spectrum (550 m), how thick a coating of MgF_2 (n = 1.38) should be vacuum-coated on a glass surface?

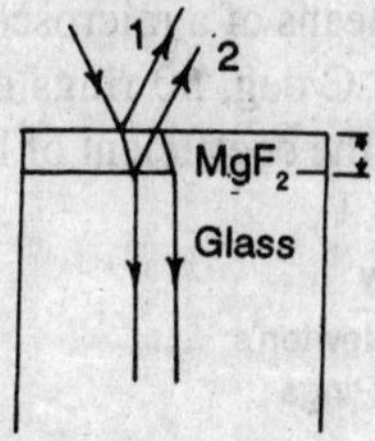

Ray striking coated glass

Fig.

Solution: Consider light to be incident at near-normal incidence. We wish to cause destructive interference between rays r_1 and r_2 so that maximum energy passes into the glass. A phase change of 1/2λ occurs in each ray, for at both the upper and lower surfaces of the MgF_2 film, the light is reflected by a medium of greater index of refraction. When striking a medium of lower index refraction, the light is reflected with no phase change. Since in this problem both ray 1 and 2 experience the same phase shift, no net change of phase is produced by these two reflections. Hence the only way a phase change can occur is if the 2 rays travel through different optical path lengths. (The optical path length is defined as the product of the geometric distance a ray travels through and the refractive index of the medium in which the ray is travelling). For destructive interference, the 2 rays must be out of phase by an odd number of half wavelengths. Therefore, the optical path difference needed for destructive interference is,

$$2nd\,(2N+1)\,\lambda/2 \qquad N = 0, 1, 2, \ldots \text{ for minima}$$

Note that 2nd is the total optical path length that the rays traverse. When N = 0,

$$d = \frac{1/2\lambda}{2n} = \frac{\lambda}{4n} = \frac{550\,m\mu}{4 \times 1.38} = 100\ \text{m}\mu$$
$$= 1.0 \times 10^{-5}\ \text{cm.}$$

Problem 10. A thin lens of long focal length is supported horizontally a short distance above the flat polished end of a steel cylinder. The cylinder is 5 cm high and its lower end is rigidly held. Newton's rings are produced between the lens and the upper end of the cylinder, using normally incident light of wavelength 6000 Å, and viewed from above by means of a microscope. When the temperature of the cylinder is raised 25 C deg, 50 rings move past, the cross-wires of the microscope. What is the coefficient of linear expansion of steel.

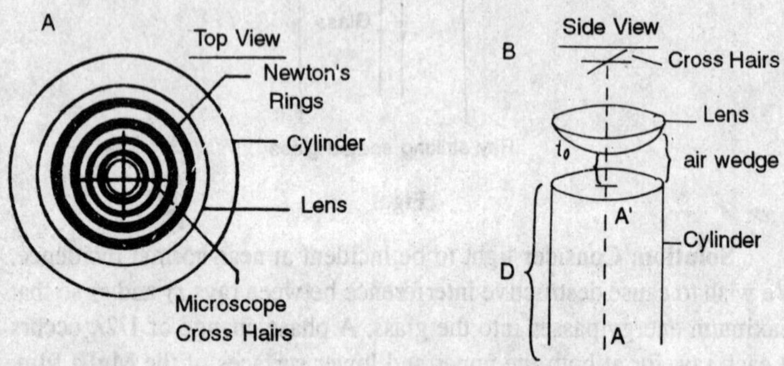

Fig.

Solution: Initially, the cylinder has a length *d*, and a gap of length To exists between the top of the cylinder and the bottom surface of the lens in the portion of the air wedge viewed in the microscope at the same position as the cross-wires. (Technically, t_0 is the distance of the lens from the cylinder as measured along axis AAé in figure (B). However, since the lens is thin, t_0 is the distance of any point on the bottom surface of the lens from the top of the cylinder). Let us now see how interference fringes are produced.

Light is incident upon the lens from above. Light is transmitted through both lens surfaces with no phase change. However, the light reflected from the bottom lens surface and the top of the cylinder are 180° (or λ/2) out of phase. This phase change is due to the reflection from the cylinder. In addition, the wave reflected from the cylinder travels a distance $2t_0$ greater than the ray reflected from the lens in traversing the air gap. If this distance is equal to an odd number of half-wavelengths (odd because the waves are already λ/2 out of phase due to reflection) constructive interference will occur. Hence

Constructive

Interference $2t_0 = (2n+1)\dfrac{\lambda}{2} \qquad (n = 0, 1, 2, \ldots)$

or $\qquad 2t_0 = (n + ½)\, l \qquad (n = 0, 1, 2, \ldots)$

Hence, for a bright fringe to appear at the crosswire, $2t_0 = (n + ½) \lambda$, where n is an unknown integer.

After one heats the cylinder through a temperature difference T, the length of the cylinder is $d (1 + \alpha T)$, where a is the coefficient of linear expansion of steel, and the gap will have been reduced to t, where $t_0 - t\ d\alpha T$.

If a bright fringe is again seen at the position of the cross-wires, then, similarly, $2t = (m + ½)\ l$, where m is an integer. $(m \neq n)$

$$\therefore\ 2\ (t_0 - t) = 2d\alpha T = (n - m)\lambda.$$

During the heating process, $(n - m)$ bright fringes must have passed over the cross-wires. Thus

$$\alpha = \frac{(n-m)\lambda}{2dT} = \frac{50 \times 6 \times 10^{-5} cm}{2 \times 5cm \times 25c\ deg}$$

$$= 1.2 \times 10^{-5} \text{ per °C}.$$

Problem 11. A radio telescope sited on the edge of a cliff overlooking the sea operates on a wavelength of 100 m. A radio star rises above the horizon and is tracked by the telescope. The first minimum of the received signal occurs when the star is 30° above the horizontal. Explain why the minimum occurs and determine the height of the cliff, assuming that radio waves suffer a phase change of π on reflection at a surface (See figure).

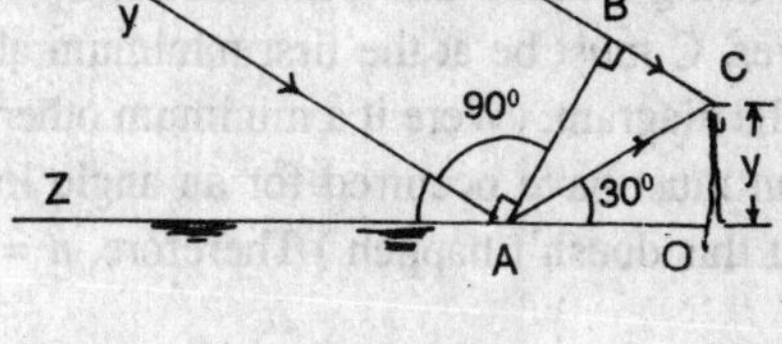

Fig.

Solution: The radio telescope is receiving signals direct from the star (ray XBC) and also by reflection from the water surface, (YAC) which is acting like a plane mirror. In this case the source of radiation is far enough away to be considered at infinity and the rays are descending on the earth in a parallel beam. When the path difference between rays YAC and XBC (see figure) is such that they reach the telescope exactly out of phase, destructive interference occurs and the signal at C will be a minimum.

Consider the diagram. If AB is drawn perpendicular to XC, it will also be perpendicular to YA, and AB is a plane wave front of the incoming beam. Thus the disturbances at A and B have the same phase at all times. A phase change of π occurs at A and this is equivalent to an increase in path of $\lambda/2$. Thus the path difference between the two rays is $\delta = AC + (\lambda/2) - BC$. But by the laws of reflection $\angle YAZ = \angle CAO = 30^\circ$, and therefore $\angle BAC = 180^\circ - 90^\circ - 30^\circ = 30^\circ$

$$\delta = AC + \frac{\lambda}{2} - AC \sin 30^\circ = \frac{\lambda}{2} + AC\,(1 - \sin 30^\circ)$$

Since sin 30° = Y/AC = Y/Sin 30° and

$$\delta = \frac{\lambda}{2} + \frac{y}{\sin 30^\circ}(1 - \sin 30^\circ) = \frac{\lambda}{2} + y.$$

In order for minima to occur, the path difference *d* must equal an odd number of half wavelengths,

of $\quad \delta = (2n + 1)\dfrac{\lambda}{2} \qquad\qquad n = 0, 1, 2, \ldots$

Then $(2n + 1)\dfrac{\lambda}{2} = \dfrac{\lambda}{2} + y$

If $n = 0$, $\delta = \lambda/2$ $y = 0$, and this corresponds to the centre of the interference pattern being a minimum. Since this is the first minimum the telescope receives, C must be at the first minimum above 0 for the situation shown in the diagram. (Were it a minimum other than the first, a previous minimum must have occurred for an angle lower than 30°. But we are told that this doesn't happen.) Therefore, $n = 1$ and δ must be $3\lambda/2$.

Hence, $\dfrac{3\lambda}{2} = \dfrac{\lambda}{2} + y \qquad$ and $y = \lambda = 100$ m.

8

Optical Diffraction and Interference

Problem 1. When light is incident on a thin slit, a diffraction pattern, shown in figure (a), is produced. Find an expression which given the location of the minima of the diffraction pattern in terms of the angle φ. (See figure (b)).

Solution: In this problem we consider only Fraunhofer

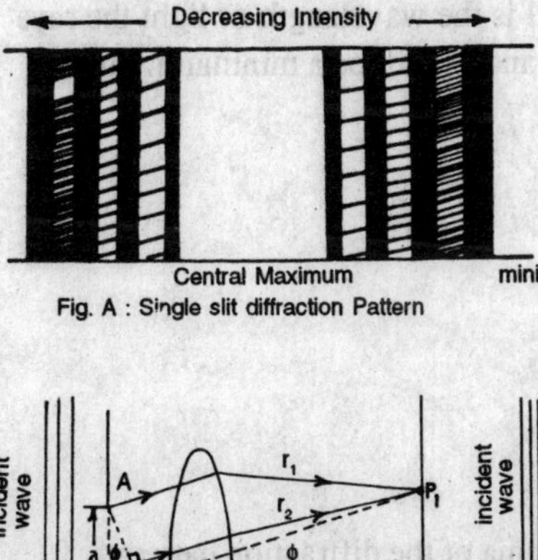

Fig. A : Single slit diffraction Pattern

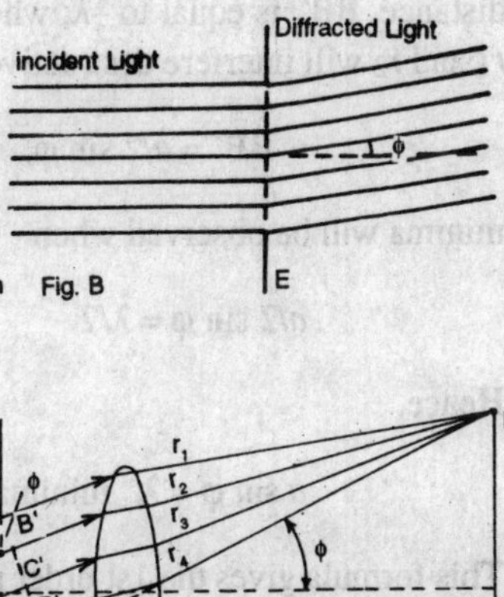

Fig. B

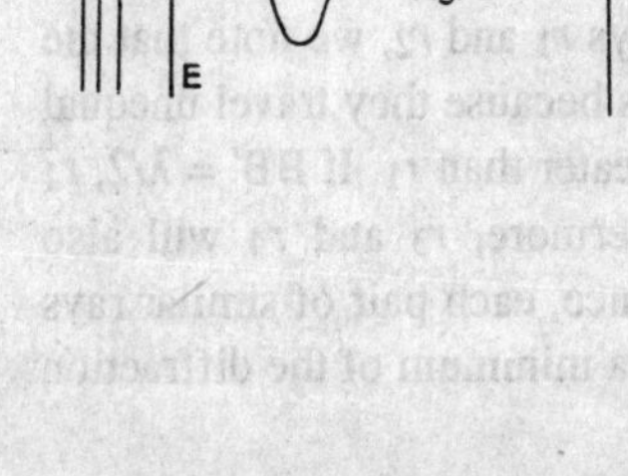

Fig. C

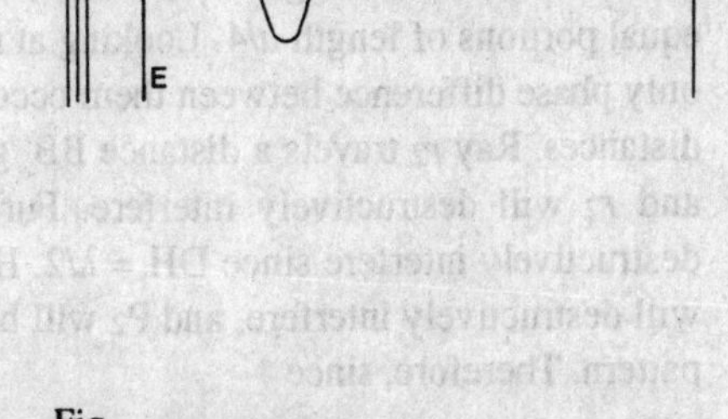

Fig. D

Fig.

diffraction. (See figure (b)). This type of diffraction is characterized by parallel, incident light rays encountering a slit, and being diffracted, again parallel to one another. In practice, this restriction may be effected in 2 ways. The light source is placed very far away from the slit, and the screen is far from the slit. Another method employs a converging lens to the left of the slit, and a converging lens to the right. Light rays incident upon the first converging lens leaves parallel to one another, encounter the slit, and are defracted parallel to one another (see figure (c)). They then are focused by the second converging lens on screen F, and the diffraction pattern results.

Using figure (c), we can derive an equation locating the minima of the diffraction pattern. Let us focus on rays coming from points A and B. Since the incident light is in phase, the only phase difference between r_1 and r_2 must occur because r_2 travels a larger distance than r_1. (Here, we assume that the rays are effectively parallel). If this distance, BB′, is equal to $\frac{1}{2}\lambda$, where l is the wavelength of light the rays r_1 and r_2 will interfere destructively and P_1 will be a minimum. Since

$$BB' = a/2 \sin \varphi,$$

minima will be observed when

$$a/2 \sin \varphi = \lambda/2$$

Hence,

$$a \sin \varphi = \lambda \quad \text{minima}$$

This formula gives the 1st order minima of the diffraction pattern.

Now, examine figure (d), where the slit has been divided into 4 equal portions of length $a/4$. Looking at rays r_1 and r_2, we note that the only phase difference between them occurs because they travel unequal distances. Ray r_2 travels a distance BB′ greater than r_1. If BB′ = $\lambda/2$, r_1 and r_2 will destructively interfere. Furthermore, r_3 and r_4 will also destructively interfere since DH = $\lambda/2$. Hence, each pair of similar rays will destructively interfere, and P_2 will be a minimum of the diffraction pattern. Therefore, since

$$BB' = a/4 \sin \varphi$$

the second order minimum is described by

$$a/4 \sin \varphi = \lambda/2 \qquad \text{minimum}$$

Then $\quad a \sin \varphi = 2\lambda \qquad$ minimum

In general, if we divide the slit into n equal segments, where n is an even integer, we will find

$$a/n \sin \varphi = \lambda/2 \text{ minima}$$

$$\text{and } a \sin \varphi = n\lambda/2 \text{ minima n even}$$

Since $n = 2m$ where $m = 1, 2, \ldots$ we obtain

$$a \sin \varphi = m\lambda \qquad \text{minima}$$

$$m = 1, 2, \ldots$$

Note that n must be even because each pair of rays emanating from the slit interfere.

Problem 2. A slit of width a is placed in front of a lens of focal length 50 cm and is illuminated normally with light of wavelength 5.89 $\times 10^{-5}$ cm. The first minima on either side of the central maximum of the diffraction pattern observed in the focal plane of the lens are separated by 0.20 cm. What is the value of a?

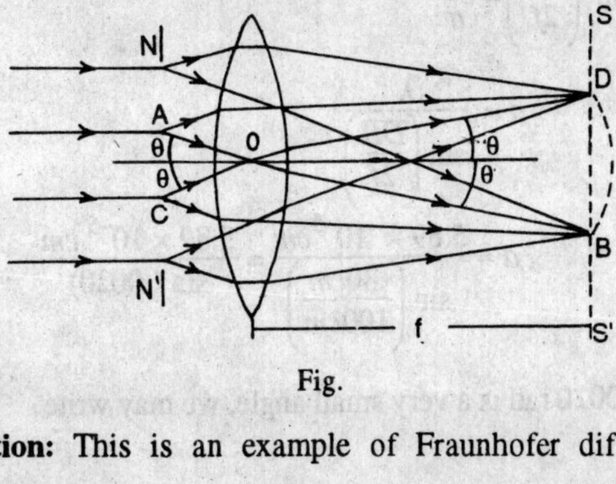

Fig.

Solution: This is an example of Fraunhofer diffraction (see figure). Parallel rays of light are incident on slit NN′ from the left. The rays are diffracted and encounter the lens of focal length f. The rays are then focused on a screen (SS′) lying in the lense focal plane, and a diffraction pattern is observed. The minima of this diffraction pattern are described by the formula

$$\sin\theta = \frac{m\lambda}{a}$$

where θ locates the minima fringes, a is the slit width, m is the minima number, and λ is the wavelength of the light used. In this problem, the first minima on either side of the central maximum is found at angle θ, such that

$$\theta = \sin^{-1}\frac{\lambda}{a}.$$

The angular separation of the 2 minima is then (see figure)

$$\delta = 2\theta = 2\sin^{-1}\frac{\lambda}{a} \tag{1}$$

But, we may write

$$2\theta \approx \frac{DB}{f} \tag{2}$$

where DB is the linear separation of the 2 minima. Here, we have approximated the arc DB (shown dotted in the figure) by the linear distance $\overline{DB}$. Using (2) in (1)

$$\frac{DB}{f} \approx 2\sin^{-1}\frac{\lambda}{a}$$

or $$\sin\left(\frac{DB}{2f}\right) \approx \frac{\lambda}{a}$$

and $$a \approx \frac{\lambda}{\sin\left(\frac{DB}{2f}\right)}$$

Hence $$a \approx \frac{5.89\times 10^{-5}cm}{\sin\left(\frac{.30cm}{100cm}\right)} = \frac{5.89\times 10^{-5}\ cm}{\sin(.0020)}$$

Since .0020 rad is a very small angle, we may write

$$\sin(.0020) \approx .0020 = 2 \times 10^{-3}$$

$$\text{whence } a \approx \frac{5.89 \times 10^{-5} cm}{2 \times 10^{-3}} = 2.945 \times 10^{-2} \text{ cm}$$

Problem 3. A plane wave of monochromatic light of wavelength 5893Å passes through a slit 0.500 mm wide and forms a diffraction pattern on a screen 1.00 m away from the slit and parallel to it. Compute the separation of the first dark bands on either side of the central bright band, that is, the width of the central bright band.

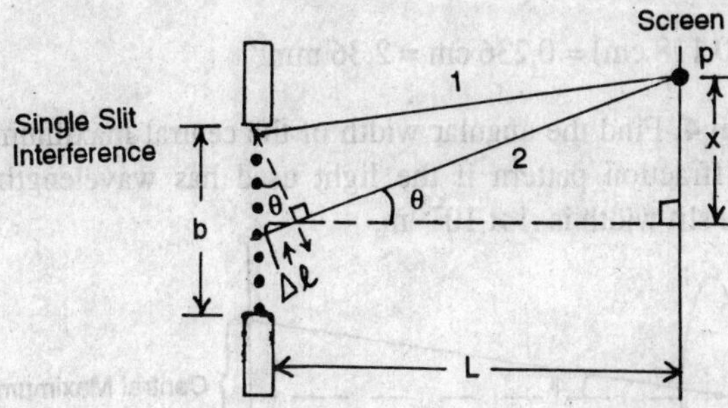

Fig,

Solution: According to Huygenes' principle, each point in the slit may be considered to behave as a single point source of light. Consider a pair of corresponding points in each half of the slit as shown in the figure. For a dark band to occur (corresponding to destructive interference) at P, the two rays must be out of phase by 180° or λ/2. This phase difference results because ray 2 travels a distance Δl, greater than ray 1 (see figure). Therefore, since P is a point of destructive interference

$$\Delta l = \lambda/2 = \frac{b}{2} \sin\theta$$

$$\lambda = b \sin\theta.$$

If L is much greater than b, we can make the following approximations: The hypotenuse of the large right triangle is equal to L, and in considering any two corresponding points in the two halves of the slit, the angle θ remains unchanged. Then, we can write

$$\sin\theta = x/L$$

But $\sin q = \frac{\lambda}{b}$ and $\frac{x}{L} = \frac{\lambda}{b}$ or

$$x = \frac{\lambda L}{b} = \frac{(0.00005893cm)\,(10)}{0.0500cm} = 0.118 \text{ cm}$$

where x is the distance of the first dark band from the middle of the central bright band. The separation of the two first-order dark bands is

$$2x = 2\,(0.118 \text{ cm}) = 0.236 \text{ cm} = 2.36 \text{ mm}.$$

Problem 4. Find the angular width of the central maximum of a Fraunhofer diffraction pattern if the light used has wavelength λ = 6000Å and the slit width is $.1 \times 10^{-3}$ m.

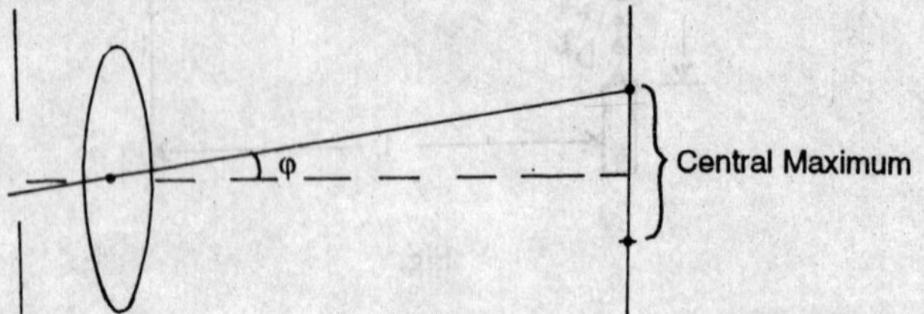

Fig.

Solution: The angular location of the minima of a Fraunhofer diffraction pattern is given by

$$\sin\varphi = m\lambda/a \qquad (m = 1, 2, \ldots)$$

where a is the slit width. The central maximum is flanked on either side by the first order minima (m = 1). The angular position of this minimum is

$$\sin\varphi = \frac{(1)\,(6000\text{Å})}{(.1\times 10^{-3}m)}$$

Since $1\text{Å} = 10^{-10}$ m

$$\sin\varphi = \frac{6\times 10^{3}\times 10^{-10}}{1\times 10^{-4}m}$$

$$\sin \varphi = 6 \times 10^{-3} = .006$$

$$\varphi = 1/3^\circ$$

Hence, the angular breadth of the central maximum is (see figure)

$$2\varphi = 2/3^\circ.$$

Problem 5. Sodium yellow light, which consists of the two wavelengths 5890 Å and 5896 Å, falls normally on a plane diffraction grating with 1500 rulings per centimetre. What is the angular separation of the two lines observed in the first-order spectrum, and under what conditions will they be seen as separated?

Solution: The grating formula is a sin $\theta = m\lambda$ where m is the order number, λ is the wavelength of the light incident on the grating, and a is the grating spacing. The angle θ locates the diffraction maxima. But the number of rulings per unit distance p is equal to $1/a$. Therefore $\sin q = m\lambda/a = m\pi\lambda$. In this problem, a small change in wavelength to $\lambda + d\lambda$ produces a change in the angle of diffraction to $\theta + \lambda\theta$. Then

$$\frac{d}{d\lambda}(\sin \theta) = \frac{d}{d\lambda}(mp\lambda)$$

or $$\cos \theta \frac{d\theta}{d\lambda} = mp$$

whence $$\theta = \frac{mp\, d\lambda}{\cos\theta} = \frac{mp\, d\lambda}{\sqrt{1 - m^2p^2\lambda^2}}$$

Note that we've used the fact that

$$\cos \theta = \sqrt{1 - \sin^2 \theta} = \sqrt{1 - m^2p^2\lambda^2}$$

The separation of the sodium lines in the first order is thus

$$d\theta = \frac{1 \times 1500\, cm^{-1} \times 6 \times 10^{-8} cm}{\sqrt{1 - 1^2 \times 1500^2 cm^{-2} \times 5893^2 \times 10^{-1}6\, cm^2}}$$

$$= \frac{9 \times 10^{-5}}{0.9961} = 9.04 \times 10^{-5} \text{ rad.}$$

We have used the average value of the 2 wavelengths of sodium light in this formula (5890 Å + 5896 Å/2 = 5893 Å).

The resolving power of a grating $R = \lambda/d\lambda = mn$, where n is the total number of lines on the grating. If we are to resolve the given wavelength difference $d\lambda$ in the first order, we must have

$$n = \frac{\lambda}{m d\lambda}$$

$$n = \frac{5893 \times 10^{-8} cm}{1 \times 6\ 10^{-8} cm} = 982 \text{ rulings.}$$

As long as the grating is wide enough to contain 982 rulings, the sodium yellow lines will be resolved in the first order.

Problem 6. The deviation of the second-order diffracted image formed by an optical grating having 5000 lines/cm is 32°. Calculate the wavelength of the light used.

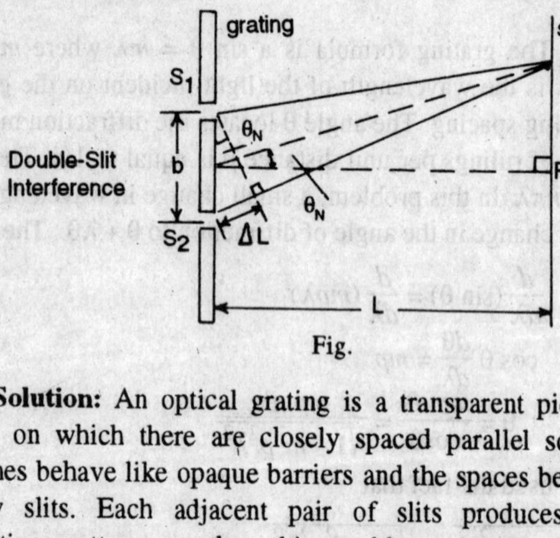

Fig.

Solution: An optical grating is a transparent piece of glass or plastic on which there are closely spaced parallel scratches. These scratches behave like opaque barriers and the spaces between slits like narrow slits. Each adjacent pair of slits produces a double-slit diffraction pattern so that this problem can be analyzed using double-slit interference.

Point P_n in the diagram receives light from both slits. The path lengths from slits S_1 and S_2 to P_n differ by an amount very close to ΔL. If the two waves are in phase as they leave slits S_1 and S_2, then they will also be in phase at P_n if ΔL is an integral multiple of the wavelength of the light. That is, if $\Delta L = n\lambda$ then reinforcement occurs at P_n and a bright band occurs there. If b is small compared to L, then the angle q in the small right triangle is nearly equal to θ in the larger right triangle. Therefore, we can say

$$\sin \theta_N = \frac{N\lambda}{b}$$

or $$b \sin \theta_N = N\lambda$$

where θ_N is the deviation of the Nth order diffracted image and b is the distance between slits.

$b = 1/5000$ cm = 0.00020 cm

$$\lambda = \frac{b \sin \theta_N}{N} = \frac{0.00020cm \times 0.53}{2}$$

= 0.000053 cm = 5300 Å.

Problem 7. The spectrum of a particular light source consists of lines and bands stretching from a wavelength of 5.0×10^{-5} cm to 7.5×10^{-5} cm. When a diffraction grating is illuminated normally with this light it is found that two adjacent spectra formed just overlap, the junction of the two spectra occurring at an angle of 45°. How many lines per centimetre are ruled on the grating?

Solution: The grating formula is

$$d \sin \theta = n\lambda$$

where λ is the wavelength of light incident upon the grating, d is the grating spacing, n is the order number, and q locates the maxima of the diffraction pattern. At the angle of 45°, we have $d \sin 45^\circ = m \times 7.5 \times 10^{-5}$ cm, and also $d \sin 45^\circ = (m + 1) \times 5.0 \times 10^{-5}$ cm. (We can see why the smaller wavelength has the larger order number by examining the grating formula, $d \sin \theta = n\lambda$. Since θ and d are the same for both l's, we obtain $n\lambda$ = const. Hence, at a particular θ, the larger the wavelength, the smaller must be n, and vice-versa).

$$\therefore \frac{m+1}{m} = \frac{7.5}{5.0} = \frac{3}{2} . \qquad \therefore m = 2 .$$

The second-order spectrum thus just overlaps with the third. Also, using the first formula above,

$$d = \frac{2 \times 7.5 \times 10^{-}5cm}{\sin 45^\circ} = 2.12 \times 10^{-4} \text{ cm}.$$

This is the separation of the rulings. Hence the number of rulings per centimetre, n, is

$$n = \frac{1}{d} = \frac{10^4}{2.12cm} = 4715 \text{ per cm.}$$

Problem 8. The limits of the visible spectrum are approximately 400 nm to 700 nm. (a) Find the angular breadth of the first-order visible spectrum produced by a plane grating having 15,000 lines per inch, when light is incident normally on the grating. (b) Show that the violet of the third-order spectrum overlaps the red of the second-order spectrum:

Solution: (a) The diffraction grating, shown in Fig. 1 has a grating spacing d given by

$$d = \frac{2.54\ cm/in.}{15{,}000\ lines/in.} = 1.69 \times 10^{-4} \text{ cm.}$$

The nth order angular deviation for wavelength λ is

$$\sin\theta = \frac{n\lambda}{d}$$

therefore the angular deviation of the violet is

$$\sin\theta v = \frac{\lambda_v}{d} n = \frac{4\times 10^{-5} cm}{1.69\times 10^{-4} cm} = 0.237,$$

$$\theta v = 13^\circ 40'.$$

The angular deviation of the red is

$$\sin\theta_R = \frac{R}{d} n = \frac{7\times 10^{-5} cm}{1.69\times 10^{-4} cm} = 0.415,$$

$$\theta_R = 24^\circ\ 30',$$

The first order deviations ($n = 1$) for the highest and lowest frequencies (violet and red respectively) of the visible spectrum will have an angular difference $\theta_R - \theta_v$. Hence, the first order visible spectrum includes an angle of

$$\theta_R - \theta_v = 24^\circ\ 30' - 13^\circ\ 40' = 10^\circ\ 50'.$$

(b) For $n = 3$, violet light has the angular deviation

$$\sin \theta v = \frac{3\lambda_v}{d} = \frac{3 \times (4 \times 10^{-5} cm)}{d}$$

The second-order deviation for the red light is

$$\sin \theta_R = \frac{2\lambda_R}{d} = \frac{2 \times (7 \times 10^{-5} cm)}{d}$$

The first angle θv is smaller than the second angle θ_R, therefore whatever the grating spacing, the third order will always overlap the second, as can be seen in Fig. 2.

Problem 9. The spectrum from a hydrogen discharge tube contains the red C line and the violet F line. A parallel beam from the discharge tube is passed through a refracting prism of 60° angle with the red C light suffering minimum deviation. What are the deviations suffered by the red C light and the violet F light?

On emerging from the prism, the light is focused on a screen by an achromatic lens of focal length 30 cm. What is the separation of the C and F images on the screen? (Here $n_C = 1.604$ and $n_F = 1.620$)

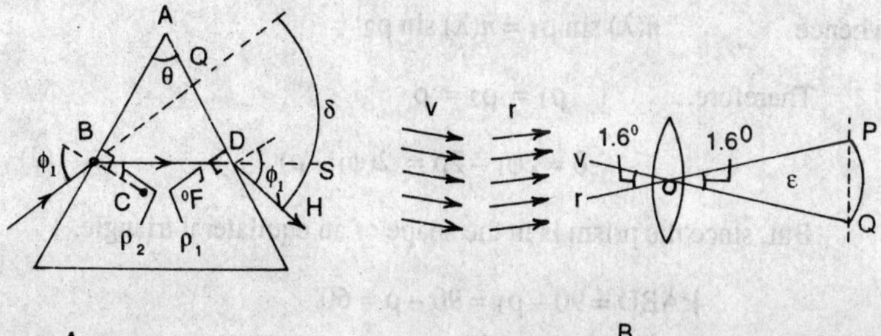

A B

Fig.

Solution: The geometry is shown in figure (a). For minimum deviation, $\varphi_1 = \varphi_2$. The physics of this problem lies in the fact that the refractive index, n of the glass of which the prism is constructed is different for different colours of light passing through the prism. That is,

$$n = n(\lambda)$$

where λ is the wavelength of the incident light. Hence, if the different coloured incident light rays are all parallel, they will leave the prism in a non-parallel manner. We now derive a formula for the deviation, *d*, of a light ray incident on the prism face at an angle φ_1.

Note, first that

$$\delta = \angle QBD + \angle SDH$$

But, $\angle SDH = \varphi_1 - \rho_2$ and $\angle QBD = \varphi_1 - \rho_1$. Then,

$$\delta = 2\,\varphi_1 - (\rho_1 + \rho_2)$$

Apply Snell's Law at both interfaces, and noting that the refractive index of air is 1, we obtain

$$\sin \varphi_1 = n(\lambda) \sin \rho_1$$

$$n(\lambda) \sin \rho_2 = \sin \varphi_1$$

Equating these 2 equations

whence $$n(\lambda) \sin \rho_1 = n(\lambda) \sin \rho_2$$

Therefore, $$\rho_1 = \rho_2 = \rho$$

$$\delta = 2\varphi_1 - 2\rho = 2(\varphi_1 - \rho) \quad (2)$$

But, since the prism is in the shape of an equilateral triangle,

$$\angle ABD = 90 - \rho_1 = 90 - \rho = 60^\circ$$

Hence, $$\rho = 30^\circ \quad (3)$$

Using this fact in (2)

$$\delta = 2(\varphi_1 - 30^\circ) \quad (4)$$

Substituting (3) in equation (1), we can find φ_1

whence $$\sin \varphi_1 = n\,(1) \sin 30^\circ = \frac{n(\lambda)}{2}$$

$$\varphi_1 = \sin^{-1}\left(\frac{n(\lambda)}{2}\right)$$

Using (5) in (4)

$$\delta = 2\left(\sin^{-1}\left[\frac{n(\lambda)}{2}\right] - 30^\circ\right.$$

We may use this formula for all the colours incident on the prism, since they all enter the latter at the same angle φ_1. Hence, for the red C lines,

$$\delta_c = 2\left(\sin^{-1}\left[\frac{n_C}{2}\right] - 30^\circ\right)$$

$$\delta_C = 2\left(\sin^{-1}\left[\frac{1.604}{2}\right] - 30^\circ\right) = 2\left(\sin^{-1}(.802) - 30^\circ\right)$$

$$\delta_C = 2\,(53.3^\circ - 30^\circ) = 46.6^\circ$$

and, for the violet F line,

$$\delta_F = 2\left(\sin^{-2}\left[\frac{n_F}{2}\right] - 30^\circ\right)$$

$$\delta_F = 2\left(\sin^{-1}\left[\frac{1.629}{2}\right] - 30^\circ\right)$$

$$\delta_F = 2\left(\sin^{-1}(.810) - 30^\circ\right)$$

$$\delta_F = 2\,(54.1^\circ - 30^\circ) = 48.2^\circ$$

A parallel beam of red light is emerging from the prism at an angle of $48.2^\circ - 46.6^\circ = 1.6^\circ$ to a similarly parallel beam of violet light. Parallel beams are brought to a focus in the focal planes of lenses, and thus both the red and violet beams give sharp images in the focal plane of the inserted lens.

The real image of a point object is the place at which all rays coverge. Hence if we can find the point at which any violet ray, and the point at which any red ray, strike the focal plane of the lens, we have located the images, and their separation gives us the answer required. The easiest rays to choose are the red and violet rays which pass through the centre of the lens, since these are undeviated.

It can be seen from figure (b) that the separation of the images is PQ. Now, if e is small, we can find its value (in radian measure), by

$$\varepsilon = \frac{PQ}{R} \tag{6}$$

where PQ is the arc PQ, and R is the radius which sweeps out PQ. If *e* is small

$$PQ \approx \overline{PQ}$$

and

$$R \approx f$$

where *f* is the focal length of the lens and $\overline{PQ}$ is the straight line PQ. This follows since both P and Q lie in the lens' focal plane. Then, using (6)

$$\overline{PQ} \approx \varepsilon f$$

where ε is in radians. Noting that

$$1^\circ = 2\pi/180 \text{ rad}$$

$$1.6^\circ = 3.2\pi/180 \text{ rad} = .028 \text{ rad}$$

and

$$\overline{PQ} \approx (.028\ rad)\ (30cm) = .84 \text{ cm.}$$

Problem 10. In double-slit Fraunhofer diffraction what is the fringe spacing on a screen 50 cm away from the slits if they are illuminated with blue light (λ = 480 nm = 4800 Å), if d = 0.10 mm, and if the slit width a = 0.02 mm? What is the linear distance from the central maximum to the first minimum of the fringe envelope?

Solution: The intensity pattern is given by $I_\theta = I_m (\cos\beta)^2 = (\frac{\sin\alpha}{\alpha})^2$, the fringe spacing being determined by the interference factor $\cos^2\beta$. The position of the first minimum of the fringe envelope is given by

$$\Delta y = \frac{\lambda D}{d}$$

where D is the distance of the screen from the slits. Substituting yields

$$\Delta_y = \frac{(480\times 10^{-9} m)\,(50\times 10^{-2} m)}{0.10\times 10^{-3} m} = 2.4\times 10^{-3}\ \text{m} = 2.4\ \text{mm}.$$

The distance to the first minimum of the envelope is determined by the diffraction factor $(\sin\alpha/\alpha)^2$ in $I_\theta = I_m(\cos\beta)^2\left(\frac{\sin\alpha}{\alpha}\right)^2$. The first minimum in this factor occurs for $\alpha = \pi$.

The angle is given by:

$$\alpha = \frac{\pi a}{\lambda}\sin\theta$$

$$\sin\theta = \frac{\alpha l}{\pi a} = \frac{\lambda}{\alpha} = \frac{480\times 10^{-9} m}{0.02\times 10^{-3} m} = 0.024.$$

This is so small that it is assumed that $\theta \simeq \sin\theta \simeq \tan\theta$, or

$$y = D\tan\theta = D\sin\theta = (50\ \text{cm})\,(0.024) = 1.2\ \text{cm}.$$

There are about ten fringes in the central peak of the fringe envelope.

Problem 11. How many fringes are formed per millimeter if light beams of wavelength 632.8 nm intersect at an angle of 5°?

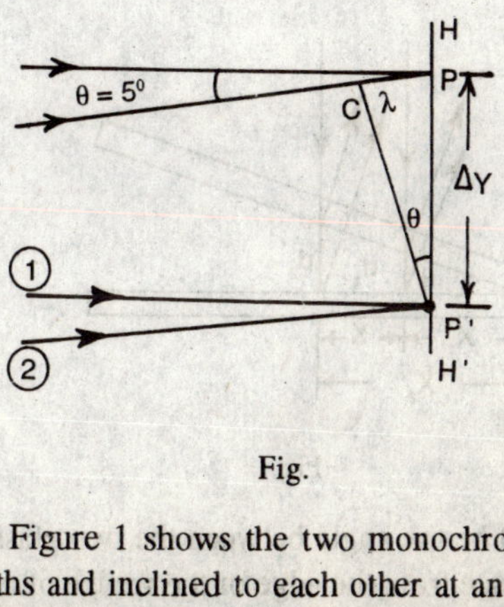

Fig.

Solution: Figure 1 shows the two monochromatic beams of the same wavelengths and inclined to each other at an angle θ. These two beams interfere on the photographic plate HH′, producing bright and dark lines. Since the beam (1) is seen to be in constant phase across the surface of the hologram plane, the interference pattern, or fringes, will be separated by an amount Δy, whenever the path difference between the two beams is one wavelength

From the triangle CPP′,

$$\sin\theta = \frac{CP'}{PP'} = \frac{\lambda}{\lambda y}.$$

Therefore,

$$\Delta y = \frac{\lambda}{\sigma\iota\nu\theta}.$$

Hence the number of fringes per mm is equal to

$$\frac{1}{\Delta y\,(mm)} = \frac{\sin\theta}{\lambda(mm)}$$

$$= \frac{\sin 5^{\circ}}{6.238 \times 10^{-4}}$$

$$= 138\ (\text{mm})^{-1}$$

Problem 12. Two glass plates are nearly in contact and make a small angle θ with each other. Show that the fringes produced by interference in the air film have a spacing equal to $\lambda/2\theta$ if the light is incident normally and has wavelength λ.

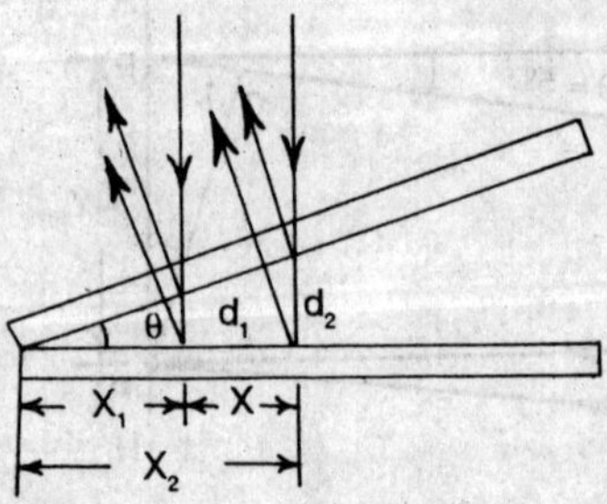

Fig.

Solution: Let θ be the angle between the two glass plates and let x_1 and x_2 be the distance from the edge of the wedge of two consecutive fringes, as shown in the figure.

If d_1 and d_2 denote the air gaps for these two fringes, then

$$2d_1 = n\lambda \tag{1}$$

and

$$2d_2 = (n+1)\lambda \tag{2}$$

where λ is the wavelength and n is an integer. Subtracting (1) from (2)

$$2(d_2 - d_1) = \lambda \tag{3}$$

From the figure, $\tan = \frac{d_1}{x_1} = \frac{d_2}{x_2}$. Assuming θ to be very small, the approximation $\tan\theta = \theta$ can be made. Therefore,

$$d_2 - d_1 = \theta(x_2 - x_1) = \theta x \tag{4}$$

where

$$x = (x_2 - x_1)$$

Substituting for $d_2 - d_1$ into equation (3),

$$2\theta x = \lambda$$

Then,

$$x = \frac{\lambda}{2\theta}.$$

Problem 13. Monochromatic light wavelength 400mμ from a distant point source falls on an opaque plate in which there is a small circular opening. As a screen is moved toward the plate from a large distance away, the Fresnel diffraction pattern on the screen first has a dark centre when the distance from plate to screen is 160 cm. Find the diameter of the central disk in the diffraction pattern if a lens of focal length 160 cm is placed just to the right of the circular opening.

Solution: We have a circular aperture illuminated by coherent light from a source infinitely distant from the aperture. This implies that all points on the aperture are in phase with all other points. We now place a screen an infinite distance from the aperture and slowly move it towards the aperture observing the centre spot of the screen. At some distance the path length from the light at the circumference of the aperture will be just one wavelength out of phase with light from the centre of the aperture, which can be represented algebraically as $R_1 = R_0 + \lambda$ in figure (2). We have learned from our study of diffraction that this is just the condition for destructive interference. Now if we move the screen closer to the aperture, the centre of the screen will have a bright spot when the path difference between a circumferential ray and a central ray is $3\lambda/2$ and dark when there is a 2λ path difference, etc. Each of the path differences defines the edges of Fresnel zones.

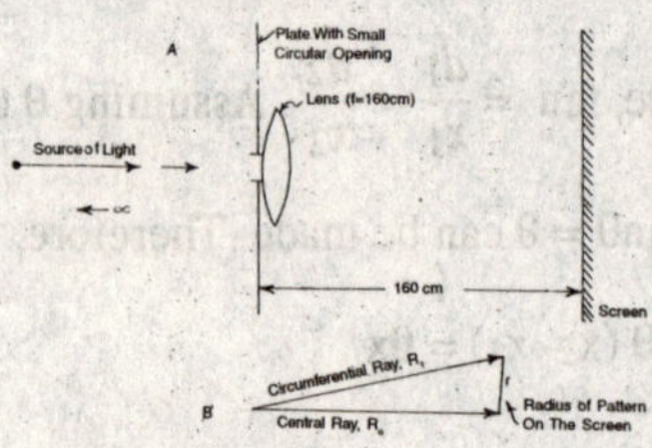

Fig.

Looking at figure (2) for our problem when R_0 is 160cm, r will just be the outer radius of the second Fresnel zone. From figure (2)

$$R_1^2 = R_0^2 + r^2 \tag{1}$$

and $\quad R_1 = R_0 + \lambda \quad$ because there is destructive interference at $R_0 = 160$ cm.

Substituting for R_1 in equation (1) yields

$$R_0^2 + 2\lambda R_0 + \lambda^2 = R_0^2 + r^2$$

The R_0^2 terms cancel, and so,

$$r^2 = 2\lambda R_0 + \lambda^2$$

or

$$r = \sqrt{\lambda^2 + 2\lambda R_0}$$

$$= \left[(400 \times 10^{-9} m)^2 + 2 \times 400 \times 10^{-9} m \times 1.6 m\right]^{1/2}$$

$$= 1.33 \times 10^{-3} \text{ m}, \tag{3}$$

which is the radius of the aperture. If we now place a 160cm focal length lens on the screen side of the aperture, the lens will convert an infinite distance on the screen side to a point 160 cm from the aperture (as in figure (1)). Stating it another way, the screen at 160 cm from the aperture with a 160 cm focal length lens will produce a Fraunhofer diffraction pattern of the source. Now from any standard optics text Fraunhofer diffraction of a circular aperture is

$$\frac{d/2}{l} = \frac{1.22\lambda}{2r} \tag{4}$$

where d is the diameter of the central disk; l is the distance from the aperture; λ is the wavelength l is the wavelength of the light used and r is the radius of the aperture. Now solving for the diameter of the central disk we have

$$d = \frac{l \times 1.22 \times \lambda}{r}$$

where $l = 1.6m$, $\lambda = 400 \times 10^{-9} m$, and $r = 1.13 \times 10^{-3}\ m$.

Hence $d = \dfrac{1.6m \times 1.22 \times 400 \times 10^{-9m}}{1.13 \times 10^{-3}m} = 6.91 \times 10^{-4}\ m.$

Problem 14. A Young's experiment is set up with the following characteristics: Monochromatic source ($\lambda = 0.55$ m), slit separation $d =$ 3.3 mm, distance from slits to screen D = 3 m (see figure).

1. Calculate the fringe spacing

2. Place a sheet of glass with plane parallel faces and thickness e

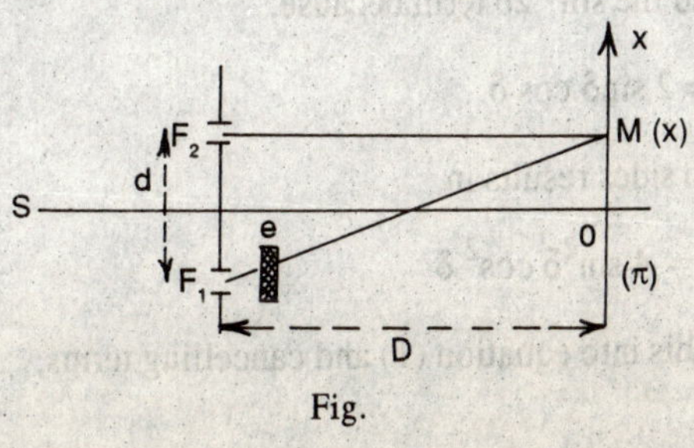

Fig.

= 0.01 mm in front on slit F_1.

(a) Determine the direction of the displacement of the fringes and the formula giving the relationship for their displacement.

(b) Knowing that the fringes are displaced by 4.73 ± .01mm find the index of refraction of the glass and its error.

Solution: Here we need to concentrate on the interference between two slits. From a standard optics text we find that the intensity of the interference pattern on N slits is,

$$I = \frac{I_0}{N^2} \frac{\sin^2 N\delta}{\sin^2\delta} \tag{1}$$

where I_0 is the initial intensity; I is the intensity at the screen; N is the number of slits; and *d* is given by,

$$d = \frac{\pi d}{\lambda} \sin \theta \tag{2}$$

where λ is the wavelength of coherent monochromatic light from the source; *d* is the slit separation, and θ is the angle between the normal to the slit system and the point on the screen. For small angles, sin $q = x/D$ where x is the distance from the centre of the screen to a point at which we wish to measure the intensity and D is the distance from slits to screen.

Substituting into equation (2) gives the result

$$\delta = \frac{\pi d}{\lambda} \frac{x}{b} \tag{3}$$

For two slits N = 2. Substituting this into equation (1) we realize we can expand the $\sin^2 2\delta$ term because,

$$\sin 2\delta = 2 \sin\delta \cos \delta. \tag{4}$$

Squaring both sides results in,

$$\sin^2 2\delta = 4 \sin^2\delta \cos^2 \delta \tag{5}$$

Substituting this into equation (1) and cancelling terms,

$$I = I_0 \cos^2 \delta \tag{6}$$

The fringe spacng i is the distance between two maxima or two minima. We therefore let $i = x^2 - x^1$ and $\delta = \pi$. Applying equation (3),

$$\pi = \frac{\pi d i}{\lambda D}$$

Solving for i,

$$i = \frac{\lambda D}{d}. \tag{8}$$

Substituting the given values for l, D, and d into equation (8),

$$i = \frac{(.55 \times 10^{-6})\,(3)}{3.3\,.\,10^{-}3} = 5 \times 10^{-4} m \tag{9}$$

Now we insert a sheet of glass of thickness 0.01 mm and index of refraction n in front of one of the slits. The light of this slits will now travel a different optical distance to a position x on the screen. Intially before the glass was inserted in front of the slit, the path difference was,

$$\delta_1 = F_1M - F_2M = x\frac{\lambda}{\iota} \tag{10}$$

The glass plate will add an additional path difference (n – 1) e so the new path difference is,

$$\delta = x\frac{\lambda}{\iota} + (n-1)\mathrm{e} \tag{11}$$

so, the new δ for a maximum will be $p\lambda$ (p being an integer) or

$$p\lambda = x\frac{\lambda}{i} + (n-1)\,e \tag{12}$$

Solving for x,

$$x = \frac{i}{\lambda}\,[p\lambda - (n-1)e] \tag{13}$$

so the shift in fringes will be the change in x:

$$\Delta x = -\frac{i}{\lambda}\,(n-1)\,\mathrm{e} \tag{14}$$

To measure the index of refraction we solve for n:

$$n = 1 - \frac{\lambda}{i}\frac{\Delta x}{e} \tag{15}$$

So we do the measurement as follows. Observe the fringe system with no glass plate, record the positions x of the maxima, then insert the glass plate, record the new positions of the maxima and determine the shift Δx of the maxima. In our problem $\Delta x = -4.73$ mm, substituting values into equation (15),

$$n = 1 - \frac{(0.55 \times 10^{-3}}{0.5} \frac{(-4.73)}{10^{-2}} = 1.5203. \qquad (16)$$

To determine the error in n we want to look at the relationship $d(\Delta x)/\Delta x$. Differentiating equation (15) we have

$$dn = -\frac{\lambda}{\iota\varepsilon} d(\Delta x) \qquad (17)$$

Solving equation (14) for $-\frac{\lambda}{\iota\varepsilon}$,

$$-\frac{\lambda}{ie} = \frac{n-1}{\Delta x} \qquad (18)$$

Substituting for $-\frac{\lambda}{ie}$ in equation (17),

$$dn = (n-1)\frac{d(\Delta x)}{\Delta x}$$

Substituting values

$$dn = 0.5\frac{2 \times 10^{-2}}{4.73} = 2 \times 10^{-3},$$

so $$n = 1.520 \pm 0.002.$$

Problem 15. Calculate, approximately, the relative intensities of the secondary maxima in the single-slit Fraunhofer diffraction pattern.

Solution: The secondary maxima lie approximately halfway between the minima and are found from

$$a = (m + \tfrac{1}{2})\,p \qquad m = 1, 2, 3, \ldots .$$

Substituting into the formula for intensity in single-slit diffraction yields

$$I_\theta = I_m \left[\frac{\sin(m + \frac{1}{2})\pi}{(m + \frac{1}{2})\pi}\right]^2,$$

which reduces to

$$\frac{I_\theta}{I_m} = \frac{1}{(m =+ ½)^2 \pi^2}$$

This yields, for $m = 1, 2, 3, \ldots$, $I_\theta / I_m = 0.045, 0.016, 0.0083$, etc. The successive maxima decrease rapidly in intensity.

9

Polarization

Problem1. Describe the polarization state of the following waves:

(1) $\overrightarrow{E_T} = i\, E_0\, [2\pi\, (z/\lambda - vt)]$

$+ J\, E_0 \cos [2\pi\, (z/\lambda - vt)]$.

(2) $\overrightarrow{E_T} = i\, E_0 \sin [2\pi\, (z/\lambda + vt)]$.

$+ J\, E_0 \sin [2\pi\, (z/\lambda + vt - 1/8)]$.

(3) $\overrightarrow{E_T} = i\, E_0 \sin [2\pi\, (z/\lambda - vt)]$

$- J\, E_0 \sin [2\pi\, (z/\lambda - vt)]$.

Solution: (1) The two components of $\overrightarrow{E_T}$ are

$$E_x = E_0 \left[2\pi \left(\frac{z}{\lambda} - vt \right) \right], \qquad E_y = E_0 \cos \left[2\pi \left(\frac{z}{\lambda} - vt \right) \right]$$

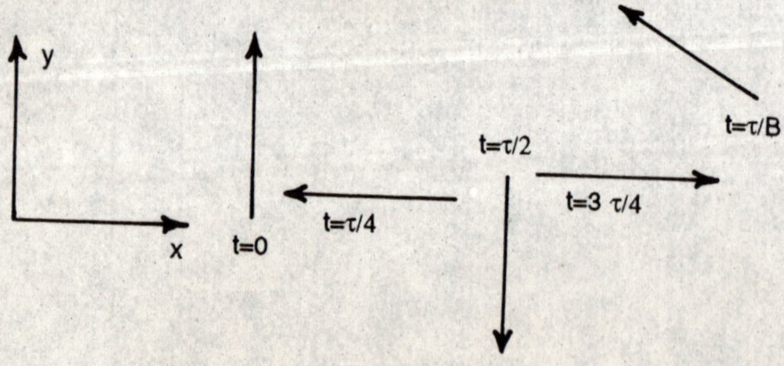

Fig.

These are actual components of a real vector, not phasors, so we want to construct the resultant at various times and also some convenient place like $z = 0$.

τ	0	$\tau/4$	$\tau/2$	$3\tau/4$	$\tau/8$	$(t = \frac{1}{\nu})$
E_x	0	$-E_0$	0	E_0	$-E_0/\sqrt{2}$	
E_y	E_0	0	$-E_0$	0	$+E_0/\sqrt{2}$	

The resultant vector is simply a vector of length E, rotating counterclockwise. The polarization is circular.

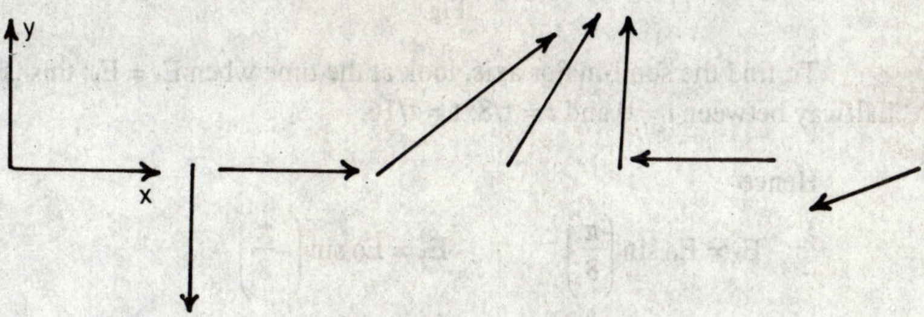

Fig.

(2) The components of the given vector are

$$E_x = E_0 \sin\left[2\pi\left(\frac{z}{\lambda} - \nu t\right)\right],$$

$$E_y = H_0 \sin\left[2\pi\left(\frac{z}{\lambda} + \nu t - \frac{1}{8}\right)\right],$$

Again we consider the point $z = 0$ and follow the resultant as time progresses

t	0	$\tau/8$	$2\tau/8$	$3\tau/8$	$4\tau/8$	$5\tau/8$	$6\tau/8$
E_x	0	$+E_0 \sin\frac{\pi}{4}$	E_0	$E_0 \sin\frac{\pi}{4}$	0	$-E_0 \sin\frac{\pi}{4}$	$-E_0$
E_y	$-E_0 \sin\frac{\pi}{4}$	0	$+E_0 \sin\frac{\pi}{4}$	E_0	$E_0 \sin\frac{\pi}{4}$	0	$-E_0 \sin\frac{\pi}{4}$

The locus of the end of $E_{Resultant}$ is an ellipse at 45 degress to the axes, as shown in figure 2b.

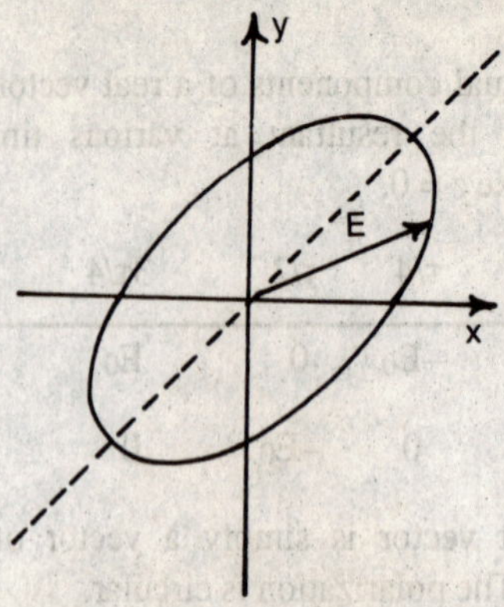

Fig.

To find the semi-major axis, look at the time when $E_x = E_y$: this is halfway between $t = 0$ and $t = \tau/8$: $t = \tau/16$.

Hence

$$E_x = E_0 \sin\left(\frac{\pi}{8}\right), \qquad E_y = E_0 \sin\left(-\frac{\pi}{8}\right)$$

$$E_R = E_0 \sin\left(\frac{\pi}{8}\right)\sqrt{2} = 0.542\ E_0$$

To find the semi-minor axis, which is $\tau/4$ later, $t = \tau/16 + \tau/4 = 5\tau/16$. So we have:

$$E_x = E_0 \sin\left(\frac{5\pi}{8}\right), \qquad E_y = E_0 \sin\left(\frac{3\pi}{8}\right)$$

$$E_R = E_0 \sin\left(\frac{3\pi}{8}\right)\sqrt{2} = 1.31\ E_0$$

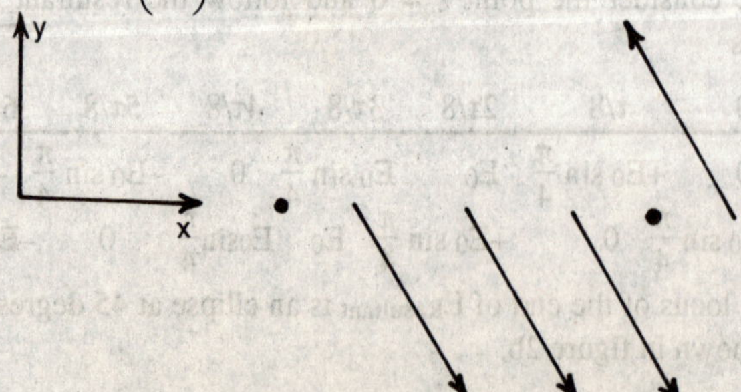

Fig.

(3) The components of $\overrightarrow{E_T}$ are

$$E_x = E_0 \sin\left[2\pi\left(\frac{x}{\lambda} - vt\right)\right]$$

$$E_y = -E_0 \sin\left[2\pi\left(\frac{x}{\lambda} - vt\right)\right]$$

As before, z = 0 is most convenient point.

t	0	$\tau/8$	$\tau/4$	$3\tau/8$	$\tau/8$	$5\tau/8$
E_x	0	$E_0 \sin\frac{\pi}{4}$	E_0	$+E_0 \sin\frac{\pi}{4}$	0	$-E_0 \sin\frac{\pi}{4}$
E_y	0	$-E_0 \sin\frac{\pi}{4}$	$-E_0$	$-E_0 \sin\frac{\pi}{4}$	0	$E_0 \sin\frac{\pi}{4}$

Thus the polarization is linear, along a line at 45 degrees to the axes. It has amplitude $E_0/\sqrt{2}$.